BIOMASS NUT PRODUCTION IN BLACK WALNUT

Exploring Management

NEIL THOMAS, PH.D.

ADJUNCT PROFESSOR
SCHOOL OF ENVIRONMENTAL SCIENCES
UNIVERSITY OF GUELPH

Neil Thomas
Visit my website at www.blackwalnuts.ca

Printed in the United States of America
First Printing: June 2010
ISBN 978-0-9865914-0-2

THIS BOOK IS DEDICATED TO MY WIFE

ANA

AND TO ALL MY ALGONQUIN COLLEGE COLLABORATORS

PAST, PRESENT AND FUTURE

TABLE OF CONTENTS

1. Introduction

I first became interested in black walnut in the 1980s when I was looking for an alternative to a livestock-based farm enterprise. Let me say right off that I had no vision of an immediate income. My own rural livelihood strategy allowed me to live on the farm but derive my income elsewhere. If I look at this issue critically, I'd say that I had the problem faced by many - what do you do with a farm if none of the traditional avenues of income generation work anymore? Most of the natural capital, in terms of original forest, is long gone. Re-establishing it would require a hundred years.

By a circuitous route, which had to do more with thinking about timber value, I came to black walnut. Black walnut is native to and therefore widely adapted throughout eastern North America, has been long cultivated in eastern Europe, and is trickling into the corners of the southern Hemisphere. It is a hardier cousin to the pecan (another native species) and to a closer relative, the Persian walnut (a more recent interloper). However, one needs the immortality of the gods to remain focused on timber.

This is not a manual. Most manuals are laid out in a very established pattern: site selection, seed selection, planting, germination, disease control, etc. This one is purposely different. A farmer neighbour, a few years ago, leaned on our fence and asked: "Is there any money in that?" I didn't know then, but I do know now and have a clearer idea of the whole process in which we've immersed ourselves, and the direction in which we're heading: Innovation. Innovation would not be served by a manual.

It should be reasonably clear by now that I'm talking about extensive planting and growing of black walnut for nuts and nut products. But while it is still important to cover all the basics, I want to cover them from a viewpoint of purpose. This purpose must be defined by you, the reader. If my neighbour reads this, he will say: "My purpose is making money. Can I make more money from that than from what I'm currently doing?" Your purpose might be different. You might say: "I have this land that I'm not using. Will black walnut for nuts bring an appreciation in value?" Or you might say: "I want to grow nut trees. What should I plant and how should I do it?"

So we start off with a range of reasons why you might want to produce nuts from black walnut. Here, I'll try to help you. I hope you'll read the whole book, whatever your reason. Before we set off, remember that this is both an experiment and a project. We are generating knowledge as we go, moulding our enterprise, and this work is not anywhere near completion. Hopefully we will revisit it five years from now and add what, in the meantime, we have discovered. If we do so, I hope it will be even more useful then. Until then I hope that what we have assembled so far will be of use to you in deciding what you might do with your land.

I have chosen the sub-title: Exploring Management" quite carefully. This is what we're doing and this is what I want to describe. This book describes a journey of exploration, and I hope that the waypoints are of more than just interest to the reader. Some may seem esoteric, but I believe an understanding of black walnut requires the net to be cast wide.

Science and Innovation

This book is written for readers challenged practically in the way I was challenged conceptually. I needed to work out a management approach adapted to what I have, and there was very little recently written which was relevant. Most modern writing focuses on a strategy which is quite different – grafted named selections, of which I have none.

Where possible, I have used scientific principles and methods. It is not peer-reviewed science for two good reasons. Firstly, it takes a significant amount of time to get scientific findings reviewed and published through scientific journals. While there is no doubt that, for scientists, this is essential for credibility, there comes a moment when one is forced to reflect on one's allotted time on this planet, and to recognize that there may not be the time to filter everything in this way. I have found that peer review often forces an outcome others wish to see. I also personally believe that much peer-review reduces innovation - and this may depend as much on the mood of the reviewer at that moment in time as it does his or her interpretation of the proponent's vision. It has been said that peer-review cannot be wholly objective:

'...specialist peer-reviewed literature can contain a lot that is wrong, trivial or misleading.'[1]

I promise I won't mislead you (not intentionally, anyway), and I shall try to avoid the traps of triviality and error.

Secondly, I decided to write the book the way I wanted to write it. I want you to understand what I think, what has happened at Lostwithiel Farm, and, where appropriate, what I have done as a result. In the same way Louis Bromfield described Malabar Farm. If he had submitted his initial plans to peer-review before he launched them, he would probably have been dragged over the coals and we should be the poorer for it (the book would not have been written, and we would have missed a major dialogue on humankind's relationship with the land). Perhaps scientists will use my observations and, through peer-reviewed literature, either accept or reject them. That is their privilege. By the way, I do not consider myself in Louis Bromfield's league - that would be vain beyond the extreme. But writers such as he and John Stewart Collis have been my models, if for no other reasons that they are so approachable, and you can feel and understand their passion.

So this book, above all, is about innovation. Modern agriculture, especially in the county where I live, is under stress. Largely commodity-driven, it is hard for a farmer to be an innovator, except when it comes to the practical use of his or her land. Improvements today in genetics and technology also happen beyond the farm gate, leaving the farmer largely as an local evaluator of cost:benefit ratios for income generation. Options are few, and those that exist are subject to shrinking margins.

This book is about biomass nut production (BNP), a term that will be explained more fully in Chapter 2. Elements of BNP already exist in the scientific literature. The innovation stems from the approach taken to test and adapt the concept - and the need to recognize this approach as both a project and an experiment. It is an experiment in that we are testing hypotheses on our data (running as we walk), but it is a project in that we are shaping it as an enterprise as we go along (making business decisions which should affect economic outcomes). From this latter perspective we require a successful outcome, but given the state of BNP art, we must define directions and take decisions as we do so. The terms 'project' and 'experiment' appear at various places in this book, and where I use them, I am not necessarily trying to make a fundamental distinction in the processes I am talking about - our project is an experiment.

The major characteristic of our process of testing, refining and analyzing is the annual cycle. Perennial species such as black walnut behave according to three principles:

- the genetics (G) of the individual tree;
- the environment (E) in which the tree grows, which can include interactions with other trees; and
- its previous or past performance (PP).

The genetics x environment (G x E) interaction is codified in the phenotype, which is the tree you actually see. However, unlike an annual plant, which grows from seed, and goes to seed, in a single year, and whose growth can be closely interpreted in the context of the environmental conditions occurring during that time, a perennial plant builds on what grew before. Years may vary, and so the phenotype is actually the cumulative expression of genetic potential in annual cycles of variable conditions. The best example of the problems we face is that of the Alternate Bearing Index (ABI), a measure of the (in)consistency of nut production year -on- year. Nut trees have a natural (i.e. genetically-based) tendency to produce variable quantities of nuts year-on-year, and the greater the variability the lower the ABI (which ranges from 0 to 1: i.e. perfect biennial bearing to uniform annual bearing). However, the measured ABI is the expression of the G x E x PP interactions, and it is known that several years of data must be collected before attempting to calculate the ABI for any given tree, and that this calculated value is only good for the site on which the tree grew (one component of environment, E) though it might serve as a vague indicator of performance elsewhere. The ABI is treated in Ch 5 .

Scientific innovation normally requires multiple experiments in order to test the hypotheses that characterize the platform upon which the innovation is based. This platform can be said to have physical and conceptual limits, and we shall describe these as we go along. Basically they are those which enable us to transform the concept of BNP into the practical farm enterprise of BNP. So most of our experimental data is derived from multiple-year observations on the variables listed below on many hundreds of trees. A partial exception to this relates to the nut variables, which, because many of our trees were still growing towards onset of bearing, were initially examined on nuts collected from trees off-farm. Many of our trees are now bearing, so this prior work has been invaluable in helping us understand what we have within the boundaries of our plantations.

4

We have also had to jump into technological innovation, because it was very clear early on that there were no suitable mechanical solutions to many of the operations we would eventually have to undertake when our trees began producing nuts. We chose to do this simultaneously in order to be prepared when this moment came. This book does not describe this aspect, because it is still ongoing. As our trees have started to produce nuts, you may think we did not plan perfectly. As not all are yet producing (nor were planned to do so simultaneously), our planning has not been too bad.

We believe that these multiple but simultaneous strategies put us in a good position to advance BNP, principally at the northern end of the species' range, though there is no reason why the principles should not apply elsewhere. Our goal is to see more nut-based agroforestry contributing to rural livelihoods.

Finally, this book is, I hope, a dialogue, where you and I can think about ideas. Many of my own ideas about longer-term components of a BNP management strategy are still ideas. By sharing them with you I am testing them in my mind and yours. Some may not stand up to the test of time (or peer review), but were I to state them as axia or rules, I would stand much greater risk.

I have tried to be consistent about certain conventions and abbreviations I've used in this book. The conventions address footnotes or endnotes, whereas the abbreviations address symbols or acronyms I've used principally to shorten the full name of tree variables, e.g. diameter at breast height, to something more easily read and remembered: DBH.

The list of abbreviations is long, so I have divided it into two: tree and nut variables. I suggest you photocopy the abbreviations table pages (Table 1-1), and laminate them back-to-back in plastic. This will form an ideal bookmark, if somewhat large, but allow you to make simple reference to the list as you work your way through the book.

Direct references to sources are generally given as chapter endnotes. I have not over-loaded the text with references.

Tree Variables		Nut variables	
Abbreviation		Abbreviation	
D	Diameter (generally of stem at specified point)	L	Distance between nut's 'poles'
DBH	Diameter at breast height	W	Distance across nut's widest equatorial axis
H	Height (total)	D	Distance across nut's narrowest equatorial axis - normally that across the face of the septum (the surface at which the two halves of the nut are connected)
HFB	Height to first branch		
DyH	Dynamic height (H-HFB)		
CSA	Cross-sectional area (generally at, and calculated from, DBH)		
DFT	Distance from tip, the distance from the measurement of D when not DBH to the tip of the tree.	NW	Nut weight
		S	Shell
		SW	Shell weight
RGR	Relative growth rate (the rate at which a variable changes relative to its existing state; good for describing change in one year relative to the outcome of previous year, appropriate for a perennial species)	SI	Shell Index (derived from SW/NEV)
		ST	Shell thickness
		SV	Shell volume
		K	Kernel
		KW	Kernel Weight
		K%	Kernel % (derived from KW/NW)
		NV	Nut volume;
Thus:		NEV	sometimes NEV, nut external volume, to distinguish it from:
RDGR	Relative diameter growth rate		
RHGR	Relative height growth rate		
		NIV	Nut internal volume
NY	Nut yield as number of nuts per sq cm of CSA		
NgD	Nutting density		

	(Total number of nuts per tree/DyH)		
ABI	Alternate Bearing Index		
ABNY	Alternate Bearing Index calculated from NY		
ABNYpop	Adjustment to Alternate Bearing (NY) calculated for a population to measure broad environmental effects on the population as a whole		
C	Carbon		

Table 1-1. Tree and nut variable abbreviations.

If a letter in parentheses such as [A] appears, it indicates the existence of an appendix containing reference data possibly of more use to the reader.

My thanks to those with whom I've discussed walnut and agroforestry issues, and whose knowledge inevitably filters through here. I am grateful for the interest and inputs of both the Leeds County Stewardship Council and the Frontenac Arch Biosphere Reserve, both being stakeholders in the outcome. Thank you also to several persons who have variously read and commented on draft material; to my wife, Ana, for her patience; and my daughter, Meghan, for getting the book together and on to the Web.

Misinterpretation, errors and omissions remain my fault.

Neil Thomas
Lostwithiel Farm
Spring 2010

1.Collins, H. 2007. Who is wearing their true colours? New Scientist, **196**, No 2631, p58. 24 November 2007.

2. How can I make money from black walnut?

That's a very good question, and when I have the full answer I'll tell you. What I can tell you now is how I think you can make money. Black walnut is one of the highest-value hardwood species. There are many anecdotes about individual trees being sold for tens of thousands of dollars, but the general truth is a little less sparkling: to be sold for a lot of money a tree has either to be cared for in a very special way or left completely alone for decades. Either way, neither you nor I will have those coins jingling in our pockets. Nor, even, our children, because it is more likely to be our grandchildren managing the farm (should they still be interested) when those trees are ready for market. And if anyone's knocked into them with a piece of machinery in the meantime, kiss the multi-million dollars goodbye.

Modern tastes have in fact increased the value of other hardwoods relative to black walnut. The Victorian tastes for dark furniture have been replaced by those for lighter woods. Black walnut has a dark, smoky colour. Today it is more likely to be used for an accent piece than for a whole suite. Even so, the world-wide supply of quality hardwood is declining, and it is extremely unlikely that well-grown black walnut will not find a ready market. But as that market will be 50+ years down the road, most of us will never know.

Then why grow it? Because, of the hardwoods, it is one of the few that provides the potential for shorter-term income. A strategic combination of that particular income stream, with other income sources, may provide the balance that a full livelihood requires. I am, of course, talking about the nuts.

The black walnut produces a highly nutritious nut. This is not the walnut you find at the local supermarket, which is generally called the Persian, Carpathian or English Walnut, though more likely grown in California or China. The black walnut is something quite different, even if it is related to the Persian walnut: a creamy, tangy nut that is valued in confectionery and other high-end culinary uses.

The trouble is that there is, as yet, no market[1], and you and I, as growers, will have to start the slow process of development on the frontier - that of defining, seeking and establishing the market. This is a component of our innovation, and while it might seem to be more project-orientated, it may at some point have to include elements of experimentation.

Can I prove that establishing a market is a possibility? Yes. The example lies over the border to the south of us, where a single company buys all the wild-grown native nuts it possibly can from pickers across a range of more than 15 states. This company would not buy hundreds of tons of unprocessed nuts annually if it had no market. Demand therefore exists. We sell hand-cracked and separated nuts locally.

We need to generate options that yield us dollars per kilogram of shelled nut, at the farm gate. It should be reasonably clear by now that I'm talking about doing it with <u>extensive</u> planting and growing of black walnut for nuts and nut products (biomass nut production, BNP).

What are the calculations that will tell me how much I can possibly make from nut production? Looking at it from a logical standpoint, they can be summed up in a series of equations that aggregate the yield components and costs to a point where it is possible to see how returns will be derived. These are listed in Table 2-1. In them, the term $=f$ implies *is a function of*, and each generally incorporates the yield components of the previous, or another, equation in the sequence. You might find it easier to read the sequence from the bottom upwards, which actually disaggregates the components

In this book I'll try to look at these points, though you may see functional constituents changing; I'll endeavour to stick to this order, but such is my love for associated knowledge that I shall not promise to avoid tangents when they arise. These equations are not the whole story, but I hope that this structure and content will make clear what we must understand and be able to do in order to consider nut production as a component of a rural livelihood strategy. I will use the left-hand term for some of my chapter headings, and the right-hand term as chapter sub-titles. Not all chapters will be dealt with in this way as there may be other topics I wish to cover, or a lack of personal experience which renders the topic as yet 'unwriteable'.

1.	Nut wt =f(kernel wt, shell wt)
2.	Number of nuts per tree =f(tree age, genetic potential)
3.	Gross nut yield per tree = f((1), (2))
4.	Gross nut yield per hectare = f((3), number of trees per hectare)
5.	Kernel % = f((1), genetic potential, growth conditions)
6.	Potential kernel yield per hectare = f((4), (5), extractability)
7.	Net kernel yield per hectare = f((6), extraction efficiency)
8.	Net marketable yield per hectare = f((7), % undiseased kernels, storage losses)
9.	Gross income from nuts = f((8), kg sold, price per kg)
10.	Net income from nuts = f((9), costs of production and management of all and any other steps not considered).

Table 2-1. Equations that aggregate black walnut yield components

For example, when I attempted to break down the yield components in black walnut, there were other questions which we shall have to answer in deciding practical approaches to nut production for rural landowners:

- how long would it be before a tree gave nuts?
- how long would it be before a plantation generated income?
- what about the timber value of nut species?
- is the timber market more or less important than the nut market?

As with all new initiatives, the majority of such questions have to be answered through trial and error and market analysis, though some may be implicit to our disaggregated equations (e.g. question 1 logically falls under equation 3). Some of this implies risk. There are some information sources that could help partially answer the questions, e.g. experimental and commercial production centres in some of the US states, but these will not satisfy all the conditions[2].

The long-term nature of nut-tree establishment and nut production, while it gives one time to find answers to questions, also constrains 'rapid responses', especially in terms of markets. On aggregate, a black walnut plantation is unlikely to produce the same yield annually and we may need production strategies to cope with this. Black walnut BNP is likely to be an inter-generational project. I cannot be certain of the probability of a successful livelihood from BNP without being the ten-to-twenty years farther down the road I need to be, especially as the experimental information is increasingly telling me that future success depends less on an orchard model (really the standard in North America) and more on the biomass model. The major distinction between the two is the number of trees on the landscape. It is only a biomass model which can respond to as-yet unidentified income streams from environmental services. It is only a biomass model which lets me think beyond the box of a low-volume, edible product, and contemplate the true richness of the non-timber products derivable from black walnut. What is a biomass model? It is thinking in multiples of thousands of trees on tens of acres (with all the complication of materials handling such scale implies), rather than the tens or, less likely, the hundreds of trees on an acre or two, which an orchardist is likely to contemplate.

Planting a tree commits one to leaving that piece of land to that tree for fifty years or more - this is not attractive to many people, especially farmers, who may, for various reasons, want a more flexible land-use pattern. So, nut production is for people who want to commit themselves to the stewardship of a property over the long term. Farmers will say: *but if I don't know whether planting trees will provide me with replacement income for the loss of cultivated land I must set aside, why should I do it?* In response I can say: firstly, you've no guarantee that corn and soybeans will be a viable source of income over the long term and diversification may reduce some of your risk; secondly, studies suggest that incorporating trees in an agricultural landscape should not necessarily result in losses over the longer term; thirdly, there is increasing interest (even if, as yet, little action) in rewarding landowners for stewardship of natural resources, because of the benefits provided to society at large.

Perhaps two quick examples will help our understanding of the type of debate. Purdue University researchers have found that net present values and economic rates of return were more favourable in agroforestry systems than from traditional agriculture and forestry[3]. Nevertheless, University of Guelph researchers surmise that black walnut and corn intercropping will return \$42 per ha per year less than annual crops alone[4]. Why the slight difference (and note my honesty in providing contrasting results)? Well, as much of what is surmised in both cases depends on models rather than existing long-term case studies, the outcome is necessarily a result of the assumptions used in developing the models.

What appears clear is the congruence of scientific, economic and social knowledge indicating that agroforestry is a sound land-use practice. Neither of these two cases included nut production as a component of the tree yield, so it might also be surmised that returns could have been higher in both cases, though some of the timber output assumptions would also require modification. Once society started to pay for off-farm benefits from agroforestry, returns to the farmer could increase further.

But let's dream a little. In Table 2-2 I've incorporated some data which allows us to understand certain dimensions of the productive process. First of all, the table is based on the biological relationship of nut yield and tree size, represented by diameter at breast height (DBH), which is itself a proxy for tree age and physiological maturity. This relationship[5] says that as a tree grows and increases in size, it will produce more nuts. Also important to us is the number of trees we require for a commercial enterprise (i.e. one which will generate more income than will be consumed in expenses, and will offer at least a partial livelihood). Table 2-2 looks at yield (kg), for 1, 10, 100 and 1000 tree enterprises, at expanding DBH from 15 to 25 cm.

Number of trees in enterprise			1	10	100	1000
			Nut Yield (kg)			
DBH (cm)						
15			1	8	78	780
20			5	45	451	4,510
25			8	82	824	8,240

Table 2-2. Relationship between DBH and nut yield, Tennessee

The formula used in Table 2-2[6] suggests that a tree produces few nuts until its DBH reaches 15 cm, when it will produce 1 kg, but that it will produce an average of 8 kg annually at a DBH of 25 cm. As the enterprise size (tree number) increases by multiples of 10, the latter yield becomes a significant amount (tonnes rather than kg).

At this point I deliberately avoid putting economic values on these data here. I do so because they were from Tennessee, and we are working far to the north where the same relationship may not hold. I confess that I have crunched some numbers of this type, and remain convinced that there is money to be made, but I think this analysis should come at the end of the book, as a conclusion, rather than at the beginning, where it would be an assumption with no immediate local basis. Incidentally, the original observations came from non-select trees encountered in the wild, and not orchard-grown named selections.

Other Products

The kernel is not the only useful part of the nut. The shell, which is hard and dense, is used as an abrasive in sand-blasting operations where a less abrasive grit is required. The pipeline and mining industries use it in various processes. Among other things, it offers potential as a filter medium in water purification processes.

Whether or not by-products are of interest, a black walnut operation will have a large volume of them, and disposal methods will be needed. The hull is easily composted (but there is too much anecdotal comment on the harm to other plants caused by black walnut hull). We believe the shell also has fuel value.

The importance of multiple income streams becomes evident if we examine gross return in another way. Figure 2-1 demonstrates the balance between returns to high kernel percentage alone, compared to returns to a combination of nut products (and, potentially, tree services, i.e. including non-kernel products, NKPs) where these other products are costed proportionally to kernel value. This examination of kernel equivalence value is important, because, as we shall see in a later chapter, in a biomass approach we take what kernel we are given, rather than planning for maximum kernel production from the outset.

If we use a range of proportional values between 0 (i.e. equivalent to no non-kernel products) up to 25%, for our kernel equivalence value of NKPs, then each line in Figure 2-1 represents the value of the NKPs for a different base K%. By means of this figure it is possible to see that a monetization of NKPs (here just considered as shell) to 25% of the kernel value in a 20% kernel line would, if added to the value of full extraction of that 20% kernel, give approximately the same return as a fully-extracted mythical 40% kernel line in which no NKPs were marketed.

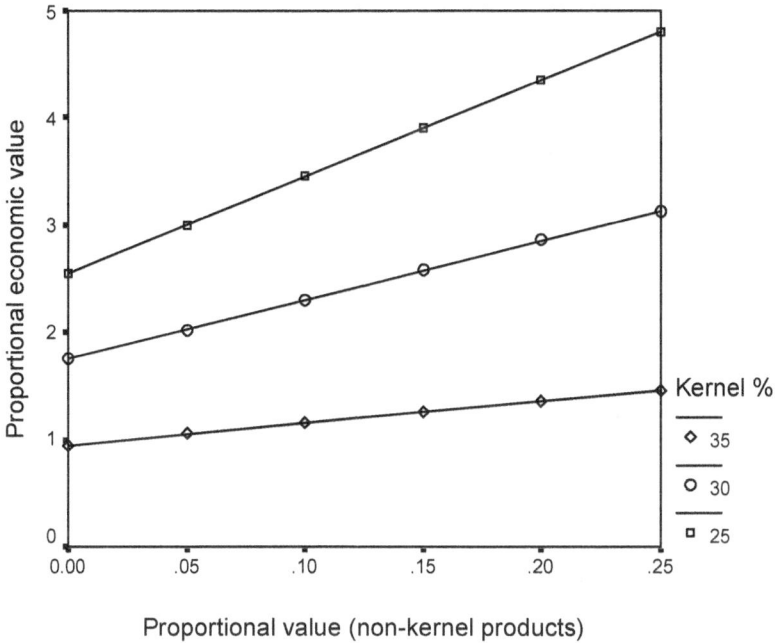

Figure 2-2. Proportional economic yield (at different kernel percentages) when non-kernel products are monetized; range corresponding to partial data from Reid et al. (2004)[7].

Lessons learned

I started this chapter wanting to convince you that there was money to be made from black walnut BNP. I'll substantiate this later, but have so far tried to give you some basis for the arguments necessary to convince you. But I also believe that the other component of the message is clear - that it will be necessary to think in multiples of 1000 trees if we are to cover costs and generate significant income. By default, then, we are launched on the biomass approach, and that is what the rest of the book will consider.

---

[1] By this I mean an economic niche which in our case might suggest that as annual 'other'-nut imports are worth the equivalent of many millions of dollars (almost $110m into Canada in 2002 from the US alone), buyers would automatically buy and/or substitute black walnut if it were available.

[2] I do not attempt to go deeply into some topics. This book is intended to spark interest; if you want to know more I suggest you read more widely, and obtain other materials, e.g. since I started writing this, new material has emerged, e.g. the Black Walnut Financial

Model at http://www.centerforagroforestry.org/profit/walnutfinancialmodel.asp

[3] Benjamin, T.J., W.L. Hoover, J.R Seifert & A.R. Gillespie. 2000. Defining competition vectors in a temperate alley cropping system in the midwestern USA. 4. The economic return of ecological knowledge. Agroforestry Systems **48** (1): 79-93.

[4] Dyack, B.J., K. Rollins & A.M. Gordon. 1999. A model to calculate ex ante the threshold value of interaction effects necessary for proposed intercropping projects to be feasible to the landowner and desirable to society. Agroforestry Systems **44** (2): 197-214.

[5] From US data. No comparable Canadian data yet exists.

[6] Derived from Brauer et al. (2006) in their reworking of 1945 and 1985 data from Tennessee: Nut-yield variations and yield-diameter relationships in open-canopy black walnut trees in southern USA. Brauer D., Ares A., Reid W., Thomas.A. and J.P. Slusher (2006). Agroforestry Systems 67 (1): 63-72

[7] Reid, W., M.V. Coggeshall and K.L. Hunt, 2004. Cultivar evaluation and development for black walnut orchards. In: Michler, C.H., P.M. Pijut, J. Van Sambeek, M.V. Coggeshall, J. Seifert, K. Woeste, and R. Overton, eds. Black walnut in a new century. Proceedings of the 6th Walnut Council research symposium, July 25-28, 2004, Lafayette, Indiana. In Gen Tech. Rep. NC-243, St Paul, MN. US Department of Agriculture, Forest Service, North Central Research Station, (188p.) 18-24.

3. Nut Weight

A function of kernel weight and shell weight

The characteristics of black walnut nuts vary principally between trees, with some variation occurring between nuts from the same trees in the same and different years. By and large, a tree's nuts do not vary a great deal in shape and size, though minor differences in linear measures can have great effects on volumetric outcomes. Here comes the first tangent.

Because I was growing impatient for my first trees to bear fruit, and because I wanted to understand what I was seeing when they did, in 2002 I began an off-farm study designed to help me. In 2002 and 2003 I collected nuts from 87 trees (16 common to the two years) across a 500km range from Niagara to Ottawa. The final number of trees from which I collected, and thus the degree of repetition of some trees in successive years, depended on the trees own fruiting pattern. In 2002, six named US and Canadian selections (the Standard: Snyder, Bicentennial, Throp, Victoria, Bowser and Fonthill) were included in the study. The objective was to determine the nut characteristics which define kernel yield in Ontario-grown black walnut. From this it was hoped to know whether locally-adapted material was a source of useful germplasm, or whether growers (including me) would have to return to the natural range and introduce new lines. How would we define a useful line? In fact, I answer this question in Chapter 5.

Fruit was processed separately for each tree in a farm-built dehuller, washed in a cement mixer, and a three-nut representative sample taken for drying and cracking. Nuts were measured before cracking, and here I must explain my own measuring system, because everything else stems from it. Figure 3-1 shows the generic model. My nut has three perpendicular values (polar, L; width, W, across the largest equatorial span; and depth, D, perpendicular across the narrowest equatorial span, generally the 'breadth' of the septum; L, W and D are measured as diameters). The fourth measure, shell thickness, or ST, cannot be measured and must be inferred independently.

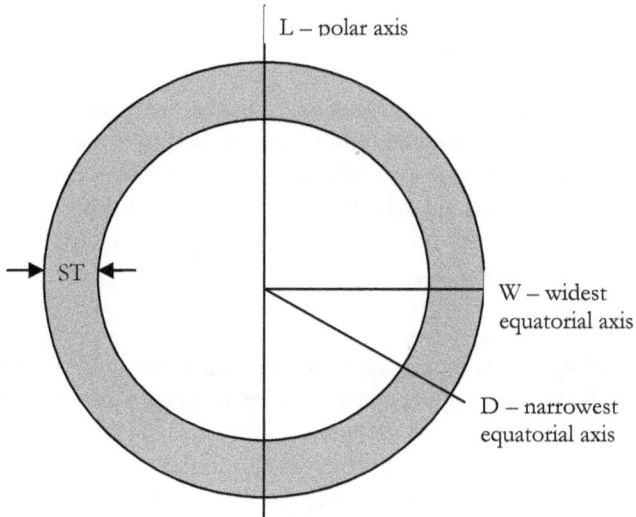

**Figure 3-1. The four critical measures in characterizing
the black walnut nut**

The characteristics of nuts collected in 2002 are indicated in Table 3-1.
Table 3-1 also shows that these ranges were well represented in the
sixteen 2003 common samples. The 2002 single-tree samples showed a
wide range in all parameters. In general, there were wild samples which
exceeded the standards in all categories. The most obvious are the ranges
for nut volume and nut weights, where wild types showed values
approximately 50% higher than the highest standard. This is less
obvious when examining data for kernel weight and kernel %,
though there were still some wild types superior to the highest
standard varieties included in the sample. In all cases there were
wild types which showed lower values than the standards.

Parameter	Range 2002 (n=81)	Mean (SE) 2002	16 Common Samples	
			Mean (SE) 2002	Mean (SE) 2003
Nut Wt (g)	28.20-9.81	15.47 (0.43)	16.85(1.35)	17.71(1.33)
Kernel Wt (g)	5.96-0.41	3.55 (0.10)	3.64(0.29)	3.44 (0.23)
Kernel %	33.36-8.16	23.08 (0.42)	21.85(1.01)	19.79(0.83)

Table 3-1. Mean Values of Nut Weight, Kernel Weight and Kernel Percentage, off-farm study, 2002 and 2003.

Visual inspection of the 2002 nuts suggested the possibility of four groupings: Kingston nuts (n=29) tended to be large and prolate, the Ottawa nuts (n=7) tended to be oblate[1], and the Niagara nuts (n=24) tended to be small; the Standards (n=6) were known to be thin-shelled, but also tended to be small. In total, these groups accounted for 66 of the 87 2002 single-tree samples. While the remaining samples were collected at points similar to, or between, the main groups, they tended to be isolated trees. The mean values for the four major parameters in these 66 samples are included in Table 3-2. The data show that the Kingston group tended to have a nut volume (NV) 30% greater than the other groups, and a nut weight (NW) 15-20% greater, though this group is not significantly different ($p=0.05$) from the Ottawa group for these parameters. While the Ottawa group showed volumes and weights slightly greater than the Niagara and Standard groups, these differences were not significant. There were no significant differences in kernel weights (KW) among groups. The Standard group had a higher mean kernel % (K%) than all other groups, though not significantly so when compared to the Ottawa group.

Parameter	N	Kingston	Niagara	Ottawa	Standard
Nut Volume (cc)	66	20.40 a	15.79 b	17.63 ab	16.15 b
	Top 25 (n)	23.83 (17)	21.80 (3)	22.94 (2)	19.53 (1)
Nut Weight (g)	66	17.59 a	13.84 b	15.18 ab	13.94 b
	Top 25 (n)	21.06 (15)	18.05 (3)	18.77 (3)	17.56 (2)
Kernel Weight (g)	66	3.76 a	3.07 a	3.87 a	3.83 a
	Top 25 (n)	4.87 (11)	4.33 (4)	4.52 (4)	5.21 (2)
Kernel Percentage	66	21.43 c	22.62 bc	25.51 ab	27.30 a
	Top 25 (n)	25.83 (5)	26.73 (7)	28.83 (5)	28.92 (5)

N.B. 66-sample means followed by the same letter are not significantly different (P=0.05)

Table 3-2. Mean Values of Main Parameters for Four Sub-Groupings, Black Walnut Study 2002

The contrasts are emphasized by examining group presence in the top 25 lines by parameter. The Kingston group stands out in all parameters except K%, indicating that even though it expressed highest overall mean KW, shell weight (SW) was a higher component of NW than in any other category. The highest overall mean K% was found in an Ottawa line (33.4%).

The inter-year relationships among the principal nut dimensions in trees common to both years, and the contrasts in the dimensional ratios are shown in Figures 3-2 and 3-3. Ascending dimensional consistency was found in the order L, W, D (R^2 values of 0.75, 0.79 and 0.82, respectively)[2]; ascending consistency in dimensional ratios occurred in the order W:D, L:W, L:D (R^2 values of 0.59, 0.84 and 0.88, respectively). There were no significant differences between years in any of these parameters (p=0.05)[3]. However, in all straightforward linear dimensions the Kingston group was significantly different (larger) from both Niagara sub-groups (p=0.05); this was not mirrored in dimensional ratios, where there was significant difference only in W:D.

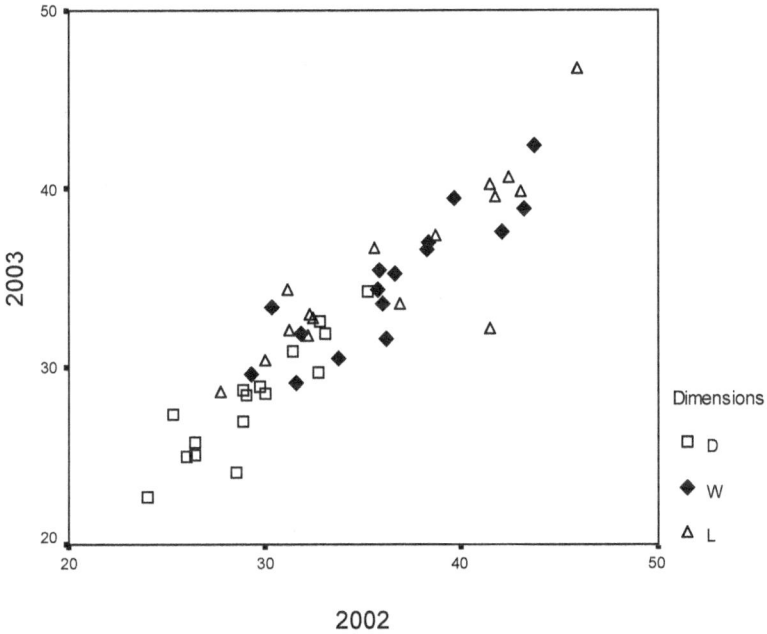

Figure 3-2. Black walnut nut dimensions: interyear comparisons of L, W and D (mm)

In spite of the lack of significant differences within linear parameters between years, it is constructive to look at three-dimensional outcomes of these parameters. Nut volume (NV, or NEV external volume) was calculated from the formula for an ellipsoid (4/3B)*(L/2*W/2*D/2). Interior (NIV) values of nut volume were calculated by subtracting from NEV the shell volume (SV, itself calculated by applying an average specific gravity of 1.3 to total shell weight).

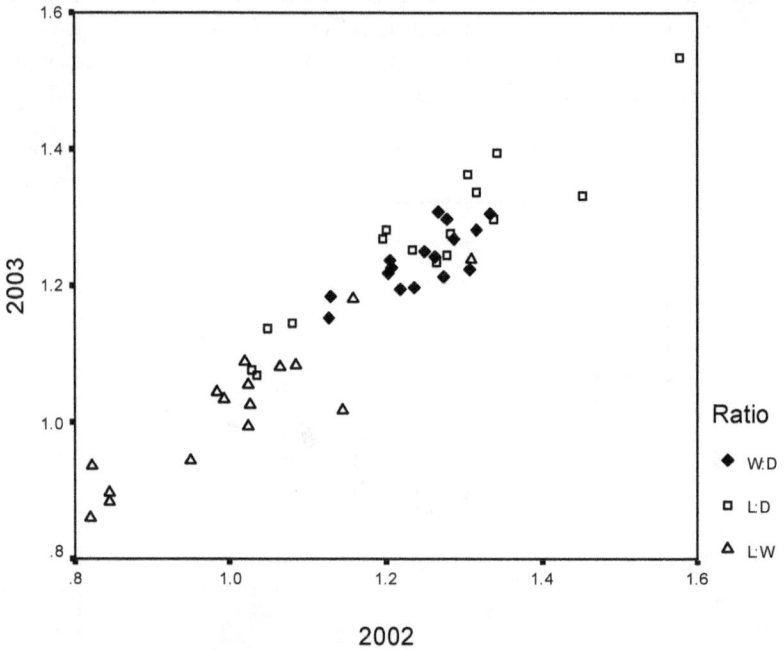

Figure 3-3. Black walnut dimensional ratios: interyear comparisons of L:W, L:D and W:D

The relationship between SW_{02} and SW_{03}, and between KW_{02} and KW_{03}, is indicated in Figure 3-4. Mean SW_{03} was 0.92g greater than mean SW_{02} ($p=0.40$); mean KW_{03} was 0.20g less than KW_{02} ($p=0.55$). The interyear within-tree consistency in SW (R^2 0.81) was higher than for KW (R^2 0.54; a quadratic function improved R^2 to 0.68). Mean $K\%_{03}$ was 2.33% less than mean $K\%_{02}$ (not shown); this is clearly evident from the difference in slope between the SW and KW comparisons. While the regression relationship is weak (best R^2 of 0.58 was given by a quadratic function), the annual mean difference ($p=0.11$) is more marked than for SW and KW.

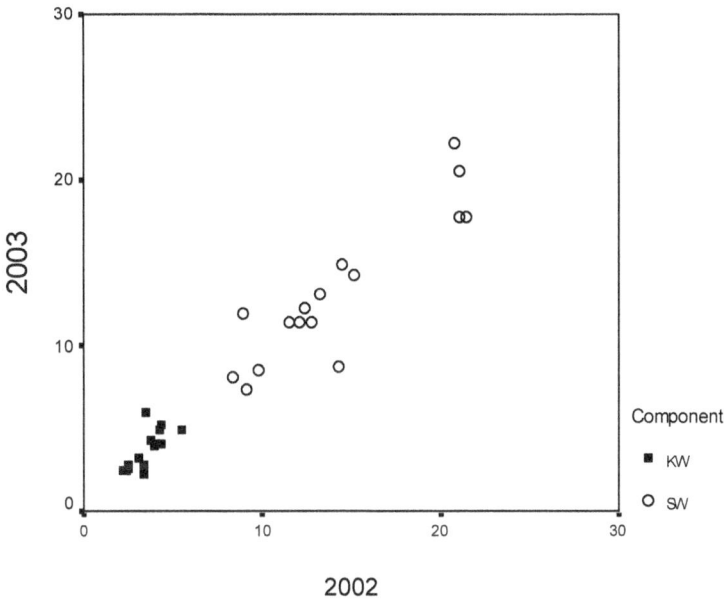

Figure 3-4. Black walnut: interyear shell and kernel weights (g)

SW, as the physical expression of NEV, defines potential kernel yield; the inter-year consistency in observed SW suggests that potential KW should be equally consistent. No measurements of shell thickness were made - the characteristics of black walnut shell render such measurement almost meaningless, and the shell involutions visible in the interior of the nut complicate this issue further. However, if shell characteristics, as SW, vary insignificantly between years, nut internal volume (NIV), i.e. the total space within the shell unoccupied by wood, is what ultimately defines both potential KW and potential K%. Calculation of NIV therefore offers an insight into the dynamics of nutfill at different nut sizes. Later I shall discuss the Shell Index, or SI, which is calculated by dividing NEV into SW, and which is the nearest approximation I think we can come to of estimating shell thickness.

In spite of the lack of significant differences between years in linear parameters and their paired ratios, mean NIV_{03} was 1.56 cc larger than mean NIV_{02} ($p=0.07$). As spherical volume increases in cubic proportion to increases in radius, a volume index for an ellipsoid will similarly be as sensitive to very small changes in its linear determinants. However, even though mean NEV_{03} was 2.28 cc larger than NEV_{02}, this difference was less significant ($p=0.16$), reflecting the smaller proportion of this difference to overall mean NEV, compared to the annual difference in NIV as a proportion of overall mean NIV. Overall mean NIV was 48.9% of overall mean NEV.

Potential KW_{03} was therefore greater than potential KW_{02}, measured in terms of possible nutfill. The plot of KW on NIV in both years is indicated in Figure 3-5. In 2002 there was a strong linear relationship (R^2 0.82), indicating that the good growing conditions had enabled trees to fill nuts uniformly, no matter what the nut size; in 2003, the larger the nut the less able the tree was to fill it. The 2003 relationship was weakly linear (R^2 0.45); transformed data using $\log_n$ values of NIV_{03} accounted for an extra 7% of the variability in the data. The quadratic function accounted for 11% more. Both functions indicate the relative decline in KW_{03} $ccNIV_{03}^{-1}$ with increasing nut size. The observed increase in NIV_{03} compared to NIV_{02} would have contributed to the lower nutfill encountered in 2003. While the annual contrast in KW was insignificant ($p=0.55$), group contrasts were highly significant ($p=0.006$). However, viewed as gKW $ccNIV^{-1}$, both annual and group contrasts were significant ($p=0.01$ and 0.04, respectively).

You may say *That was some tangent!* To which I will reply *You won't understand KW and SW as determinants of NW unless you embark on this journey with me, tangents and all.* What can we conclude from all this? The parameters L, W and D and their ratios (i.e. nut shape and size, and hence total volume) are evidently under tight genetic control. Shell thickness, expressed as SW or SI, would appear to be similarly controlled. The nut shell, as a sink for photosynthesis, is less mutable than, and appears dominant over, the kernel - it is the latter which principally expresses environmental effects on fruiting; all black walnut growers face the task of separating empty shells from full ones, evidence that at some point kernel production was switched off in the former. K% cannot be predicted from external shell dimensions or their ratios; it is the one variable that requires cracking and separation for its determination.

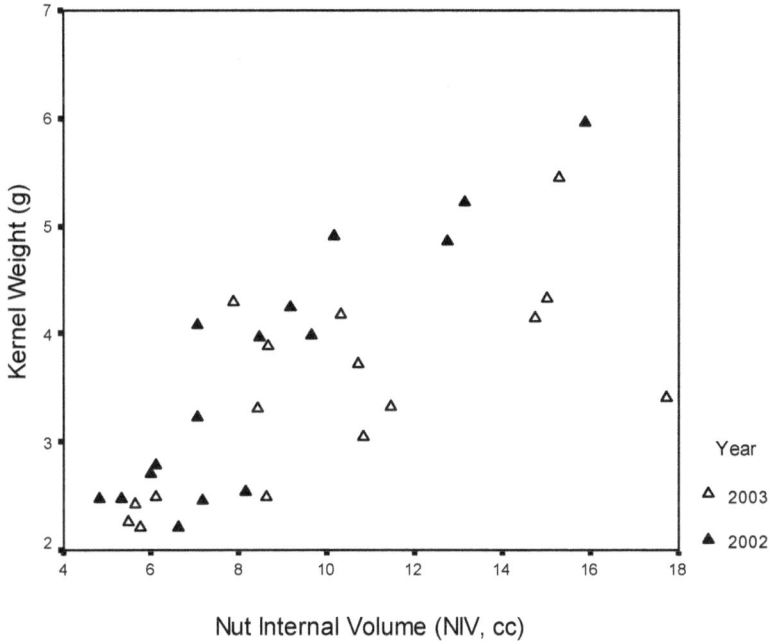

Figure 3-5. Black walnut nutfill: regression of KW on NIV

The concept of nutfill as an expression of the environment suggests a strong argument for using trees with smaller nut sizes in marginal environments, to avoid, depending on the year, a proportion of partially filled or shriveled kernels (of course, variable cluster size may offset this). Sparks[4] has also noted the increasing difficulty in filling a larger nut in pecan. However, smaller nut sizes will always give smaller kernels, and a stand thinned to only small-nutted trees will not yield possibly higher-value larger kernels in good years. High SW means high shell volume, but it is only through an understanding of its corollary, NIV, that one can interpret the dynamics of K% for any given tree in a stand. In this study, the group with highest SW also had trees with high KW, so selection for low SW is not necessarily conducive to optimizing KW. In a model 1000-tree farm-scale population, proposed by this author as the minimum definition of a biomass approach to nut production, measurements of nut size and shape may seem arduous; however, inter-year consistency in L, W, D and SW suggests that once a tree's shell characteristics are known, only KW needs to be tracked annually[5]. Niagara-grown US-named selections were variously intermediate in this study in terms of K%. Recent observations of disappointment over the cracking percentages of superior US selections[6] suggest that there is merit in not initially investing heavily in grafted stock unless growth in average annual extracted kernel yield per hectare, compared against a local control, and factoring in all the extra risks likely to be associated with such material, is proven to warrant such investment.

A further year of data was collected which allowed this analysis to be extended, but you would find it an even greater tangent so I will limit the off-farm study discussion to that presented above.

My own trees began bearing in 2004, but such was my organization, and focus on just how to manage harvest, that I didn't mimic the off-farm study until 2007. By that time I had worked out nut range scoring as a way of identifying superior trees (see ch 4), so that I could limit destructive nut analysis to the trees which were going to be top contributors.

2007

Before presenting the results, it is important to note the characteristics of the 2007 harvest. A very late spring frost (May 19th) caused widespread damage to emerging flowers. Nutting was considerably reduced when compared to 2006. Further selection of trees for harvest was on the basis of trees that actually gave a nut sample big enough to ensure sufficient nuts for on-planting. The area selected for planting of this F1 material was big enough for a randomized complete block study of 25 lines replicated 11 times. The planted sample (two nuts per location) thus consisted of nuts from the 25 maternal trees considered most interesting (low disease scores, high nut number in a low nut number year, precocious trees – nutting at a small size/young age, range in nut sizes, etc). As no cracking analysis had been done to date on-farm, there was no prior knowledge of kernel weights or percentages.

The mean nut weight (NW) was 15g, and the mean kernel weight was 2.5g. By themselves, these values don't tell much about the sample. Figure 3-6 indicates the distribution of NW and Figure 3-7 that of KW. NW ranged from about 7.6g to 18.8g. KW similarly ranged from 1.7g to 3.9g (line 1-1-17), though NW and KW are not directly correlated, i.e. lowest NW did not exhibit lowest KW, nor highest NW the highest KW.

But I've left the best to last. The sample showed a mean kernel percentage (K%) of 22%, calculated from KW as a percentage of

NW. Almost 25% of the trees sampled showed a very respectable K% of 25% or more. Go to Chapter 7 for more on this, and go to Chapter 4 for completion of the analysis of kernel yield per tree (KY, by multiplying KW by the number of nuts per tree).

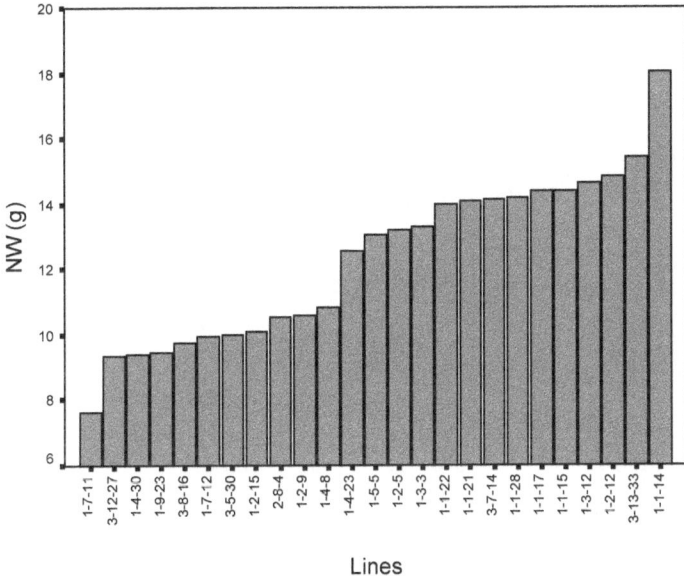

Figure 3-6. Individual NW in measured lines, 2007.

A methodological error in 2008 resulted in widespread frost damage to kernels from the top trees, so the 2007 analysis could not be repeated. However, 1.5 t of fruit was eventually gathered in 2008, confirming the increasing potential productivity of young biomass stands.

4.5

4.0

3.5

KW (g) 3.0

2.5

2.0

1.5

1.0

1-7-11 1-7-12 1-2-15 1-4-30 3-8-16 1-2-9 2-8-4 1-5-5 3-12-27 1-4-8 3-5-30 1-1-17 1-9-23 1-1-22 1-2-5 1-1-15 1-1-21 1-1-28 1-3-3 1-4-23 3-7-14 3-13-33 1-3-12 1-1-14 1-2-12

Lines

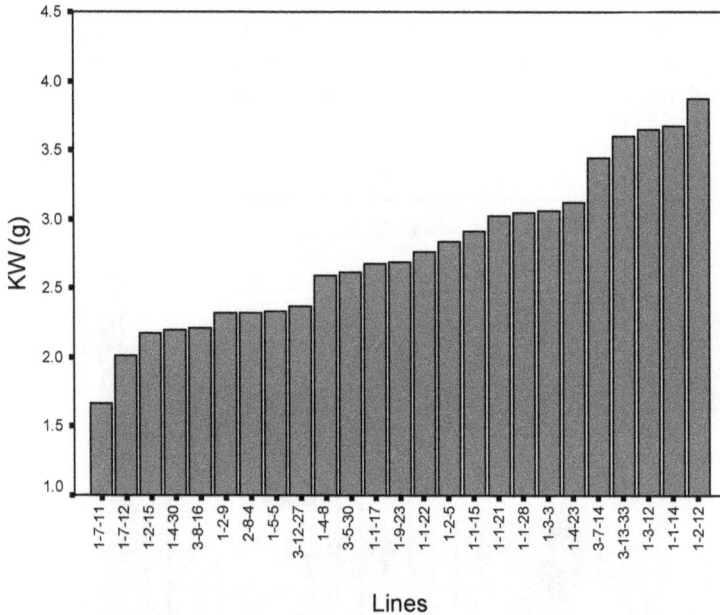

Figure 3-7. Individual KW in measured lines, 2007

Lesson learned

You should have captured that NW is comprised of two indices, SW and KW, neither of which will give you an understanding of K%. It is important to remember that high K% is generally not linked to high KW, even though the former, with an associated presumed ease of cracking, has been the driving variable in naming US selections.

I asked at the beginning: How would we define a useful line? The answer to this has to be 'replicability in those characteristics that make it productive". We have answered the question about the possibility of finding useful germplasm, and even some which had some nut characteristics equal to or better than named US selections. We lack the years of data necessary to test replicability in all aspects.

Nut genetics

Here I want to extend the discussion about L, W and D, and implications for selection. I have suggested that each variable, and SI, is under tight genetic control. This implies that there is less environmental impact on these observed shell values than there is on KW, and that within-tree inter-year expression of these variables will be only marginally affected by 'environmental noise', though we have seen that minor variation in linear dimensions can have significant effects on volumetric outcomes. What is of interest here is the issue of heritability, and what a daughter's nuts will look like in relation to the maternal line. As we don't know the paternal genetics in any line, nothing relating to the paternal line can be inferred.

We can construct a model to examine maternal genetics in more detail. If we take the largest and smallest nuts in our population and say that these represent the limits of linear changes in our variables as tested through countless generations of evolution, we can use these measures to put limits upon change in the next generation. Then we can say that if there is no heritability involved, i.e. that a daughter's nuts will not mirror the maternal line's nuts in any way, our variable changes within these limits will be random over generations. If there is some heritability (which we have not yet been able to test in the field, though the trees are in the ground), some variable change should be non-random. In either case, we need to keep in mind certain laws: D is never greater than W or L, though the latter two can vary considerably and we need to test the ratio L:W before we generate D; we also need to generate our estimate of shell thickness, SI, and we can do it in the same way as for L and W.

The ascending inter-year consistency in dimensional ratios (which occurred in the order W:D, L:W, L:D) is not evidence for heritability in linear dimensions, because it was measured on the same trees, not mother/daughter pairs. It speaks to variability between dimensions in environmental responses, with some being less sensitive than others.

If shell dimensions were random between generations, within the limits mentioned above, they might produce a result[7] like that shown in Figure 3-8 which shows computed NEV and SI (Chapter 8; *10 to place it on the same vertical axis) over 50 generations.

Here is where I will speculate. Generational 'randomness' is unlikely, because it is an evolutionary vacuum. I think time will prove that there is some maternal dominance, however little, in one or more of our four shell dimensions. If each dimension is under the control of a single gene, then fairly simple Mendelian outcomes should be in order. If several genes control each dimension, it will be harder to partition results to individual genes and any model will be more complex. One thing is clear: that it is almost certainly a waste of time looking for heritability in kernel characteristics if shell heritability is not addressed first i.e. it is SI that defines K%.

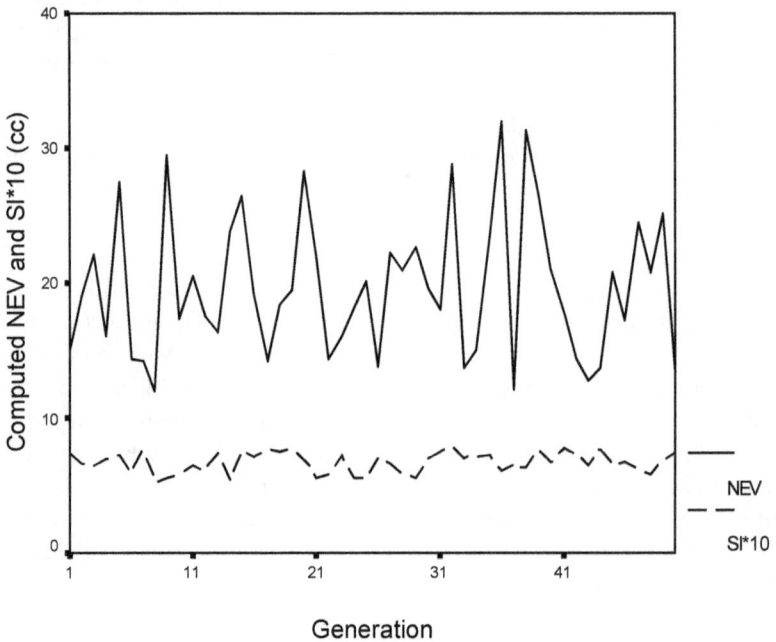

Figure 3-8. Computed NEV and SI*10 for random nut dimensional parameters across 50 generations

Another point. An examination of the energy cost of nut production suggests that in some generations an evolutionary strategy might be to test the growing conditions of the current year by pushing one or more shell dimensions to the maximum. If the energy cost can be met, the nut would have an increased probability of extending the maternal genes (or the combination of maternal and paternal genes) into the next generation; if not, the probability may be average or less. This may be the case every year, where the energy cost of formation and filling of the ellipsoid nut is measured against the current year's conditions.

[1] A prolate nut tends to be extended at the poles, an oblate nut is squashed at the poles i.e. along dimension L.

[2] An R^2 value is a measure of how much variability (maximum 1.0 or 100%) is accounted for in the relationship being described. At high values (near 1.0) there is increasing confidence that much of the variability is accounted for and that the relationship will bear scrutiny.

[3] A p value is way of expressing the probability of a difference being due to chance. A value of 0.05 suggests that in 95 cases out of 100, the observed difference will not be due to chance. Lower values, e.g. 0.01, express less likelihood of a chance result.

[4] Sparks, D.   Influence of topography, crop load and irrigation on pecan nut volume and percentage kernel and implications to production. Availability: [On-line] http://www.geocities.com/CollegePark/Campus/3370/InfluenceTopography.htm [29 June 2005]

[5] Elsewhere it is suggested that Nut Range Scores assist in highlighting those trees upon which to concentrate.

[6] Hanson, B.L., 2003. Cracking black walnuts commercially. The Nutshell; Quarterly Newsletter of the Northern Nut Growers Association. 57(3):25-26.

[7] Using Excel's© RAND function

4. Number of nuts per tree

A function of tree age and genetic potential

Counting nuts (and other things) may seem one of the more esoteric of pastimes, if not rather dull, but will be one of the most important things you do, especially if you truly want to understand individual tree productivity (you can't calculate NY without doing so). Because it is onerous for large numbers of trees, and is difficult to do when the nuts are still on the trees, we have been considering other ways of approaching the problem.

Data collection will remain a bigger issue for biomass producers than orchardists in quantifying individual tree productivity, but it will be important to both to keep good records and to understand what is happening in their stands. Nut counts will comprise a large part of this effort. One truly cannot understand individual tree productivity without doing nut counts (e.g. Jones et al., 1995). Key selection indices on which we have been working, e.g. canopy nutting density, cannot be calculated without them. Impacts of disease susceptibility cannot be understood without a systematic understanding of disease incidence throughout a plantation.

The first question we asked was: *How important is it to be extremely accurate?* The second was: *Can we use categories which represent number ranges to capture levels of productivity?* While it might seem that the easiest is to wait until the nuts are on the ground, in the long term this is unlikely to fit well with mechanical harvesting, as the nuts will not fall all at once, and we are likely to want to make frequent passes with our harvester[1]. The objective of this work has been to simplify quantitative assessments, thus reducing the amount of time physically spent in data collection and evaluation, and supporting selection methods.

Range scores (RS; nuts - NRS) were developed and tested in 2006. The tree stands are identified as Field 3 (initially established in 1994) and Field 1 (initially established in 1992). Field 3 is the sentinel site, i.e. all trees are assessed annually for all variables of interest; it is more uniform than Field 1. Field 1 was included in the test of nut counts in 2006 because of the opportunity to do so. NRS were scored in September 2006 by two two-person teams of youth-volunteer Ontario Stewardship Rangers, who also marked trees with coloured tape according to range score. NRS accuracy was assessed by the author from full counts of nuts hand-harvested from the ground in late October. Generally NRS scores were within 1% of real values. Range scores for nuts are shown in Table 4-1[2].

Score	Count
0	No nuts
1	1-5 nuts
2	6-20 nuts
3	21-80 nuts
4	>81 nuts

Table 4-1. Range scores used to quantify nut counts in 2006

The overall distribution of trees in both fields by NRS is shown in Figure 4-1. These are shown as percentages to avoid confounding the different numbers of trees in each field (504 and 578 in Fields 1 and 3, respectively). It can be seen that the pattern is very similar between fields, with large numbers of trees still non-producing, and with declining percentages in higher ranges. The high percentage of non-producing trees appears to overshadow differences between ranges 1-4.

By contrast, Figure 4-2 shows the mean number of nuts per tree by NRS. The ranges worked very well for Field 3, where the average for each range falls approximately at the mid-point for each range. The ranges chosen did not work so well for Field 1, where averages by range were almost exactly double those in Field 3. However, even if we were to increase our range sizes (e.g. '1', 1-5, '2', 6-30, '3', 31-150, '4', >150) this would hardly change the form of the bars for Field 1 in Figure 4-2 because of the overall distribution of number of nuts per tree, especially for range '4'. The new ranges would cause some trees to be placed into lower ranges, because of the increase in the lower limit of each range. Bars for '2' and '3' would increase slightly in size, while '4' would decrease slightly.

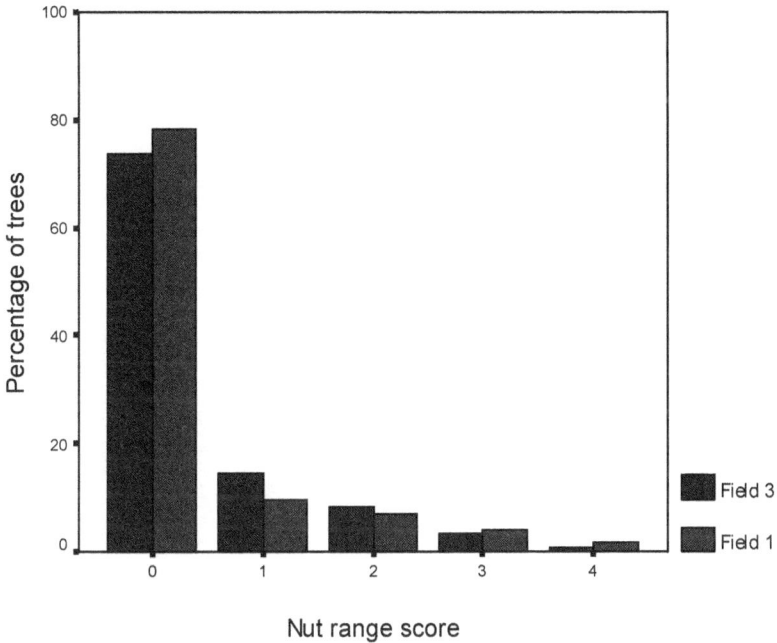

**Figure 4-1. Percentage of trees falling into different NRS,
Fields 3 and 1, Lostwithiel Farm, 2006**

What is the purpose of our numerical ranges? In the case of NRS, firstly
we want to identify those trees that have nuts as opposed to those which
don't. A simple binary code of 1 or 0, respectively, works here. However,
then perhaps we want to divide the 'with nuts' trees into further sub-
groups which indicate escalating levels of productivity.

Where did the ranges for the sub-categories come from? Frankly, from a
few scribbles on the back of an envelope while making some counts. It
was very simple to separate '0' and '1' trees. It was then quite simple to
assess whether a '1' tree had five nuts or less (remaining a '1'). It was then
not much harder to count from between 6 to 20 nuts ('2'). Category '3'
required a little more work, but once one counted up to 20, it was quite
simple to estimate visually whether total number of nuts was less than or
greater than four times this number.

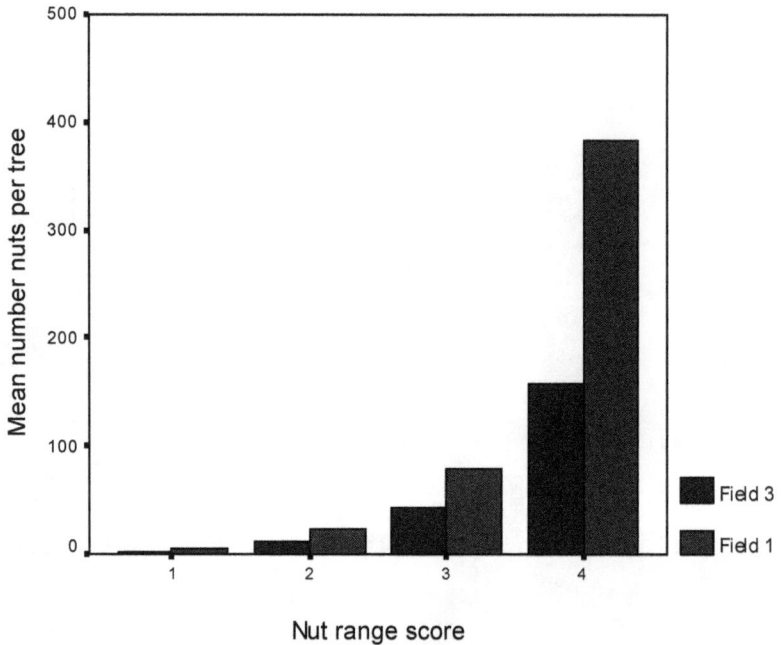

Figure 4-2. Mean number of nuts per tree by NRS, Fields 3 and 1, Lostwithiel Farm, 2006

If the estimate was 'greater than', then counting was required to ensure the number was greater than 80 ('4'). These ranges are not fixed in stone - they were more a rough estimate of what I thought would be required for our Field 3, but then we also tested them against our Field 1. The actual counts were carried out by volunteer Ontario Stewardship Rangers, who received about a ten-minute briefing on how to do it, and which colour marking tape for which range to put on the tree ('0' - none). The nut ranges fit an approximate exponential scale, where the top value of each range is greater than the lower value of the same range more or less in equal proportion.

The trees in Field 1 are different from those in Field 3. There is greater variability in ages and sizes (hence the reason why all our measurements have so far been concentrated on Field 3), with a significant proportion slightly older and larger than those in Field 3. The bars per NRS (Figure 4-1) for Field 1 are higher than those for Field 3 because overall nut counts were greater per tree on average where trees were producing nuts. The purpose in testing range scores was to examine the feasibility of avoiding a full nut count, but if this didn't actually help us to interpret the data there was little point in doing it. In our analysis of raw data (not shown) we saw only a general trend in the relationship between number of nuts per tree and DBH. However, if we plot Field 3 NRS against DBH (Figure 4-3) we immediately see that our transformation of average nut numbers per tree into NRS uncovers a clear relationship. Unfortunately we could not do this for Field 1 because DBH data were incomplete, but interpolation of Figure 4-3 suggests that a combined analysis of Fields 1 and 3 would have extended the analysis to higher DBH values. Because we have only four productive ranges, the apparent slope of the graph would have decreased - we actually needed to add a fifth range if we were to test this relationship accurately to greater DBH. Thus rather than increase range size, as considered above, we might limit '4' to 81-320 nuts, and add a '5' consisting of >320 nuts, to be consistent with our original ranges.

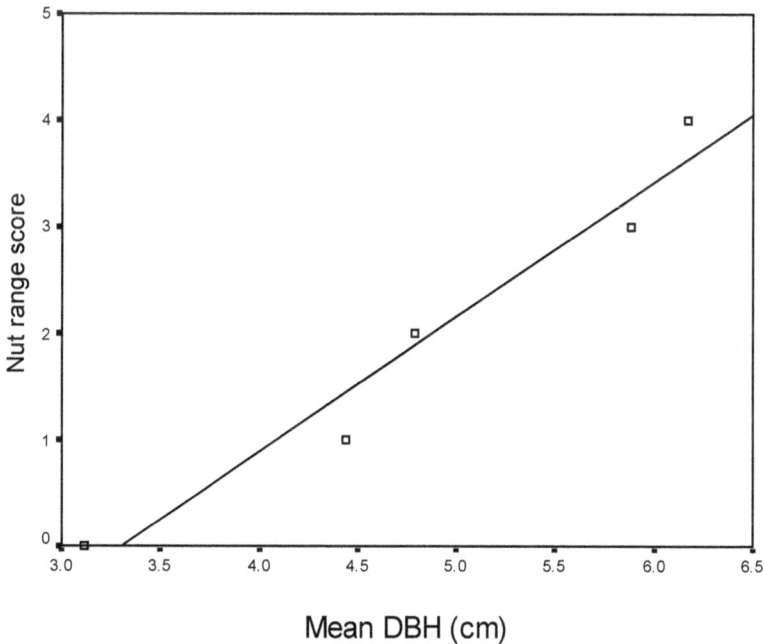

Mean DBH (cm)

Figure 4-3. NRS against mean DBH, Field 3, Lostwithiel Farm, 2006

So it appears that using NRS is actually useful, though we have seen that a fifth range, and expanded DBH measurements to another field might have helped us understand nutting onset better. The data so far suggest that significant nutting will occur at DBH below 15cm on a fair proportion on non-select trees planted. Figures 4-2 and 4-3 also indicate that nut yields increase exponentially with DBH within the ranges examined, also an important observation, because it confirms that young plantations can produce significant total yield (per acre or per ha) within a relatively short time after onset of nutting. While we had hoped to use NRS to avoid full nut counts, it is clear that adding a fifth range will increase the counting work. However, it will still help us understand nutting onset more clearly if the relationship in Figure 4-3 proves to be consistent as the DBH range expands.

It should be remembered that most of these data refer to characteristics of a population of young trees, and that they do not generally describe 'progressions' or 'tendencies' – most of them illustrate segregation. The exception is Figure 4-3, which does show a tendency of higher NRS as DBH increases. However, we need several years of data to determine the quantitative nature of the relationship (and take alternate bearing into account), as it is possible that some trees may be very late in achieving physiological maturity. The purpose in pursuing such research would lie in understanding how many non-select trees one needs to plant to be able to thin down to an economically-viable select population, and thus the original planting density one should use. Figure 4-4 shows the 2006 expression of nutting density at each NRS, suggesting that all of our NRS '3' and '4' trees are our future select individuals, but we actually need to separate DBH effects[2] from the data if we are to make this a valid decision.

At the time when we conducted this work, we expressed nutting density (NgD) as the relationship between the number of nuts per tree and canopy size. We expressed canopy size as 'dynamic height (DyH)', which was the difference between total height and clear bole length. The unit of NgD is number of nuts per cm DyH. Maximum 2006 NgD values for the top three trees in Field 3 and 5 were 0.39-0.74, and 1.03-1.42 nuts per cm DyH, respectively. We thought that NgD would be a crucial selection index, and use of NRS would allow us to identify easily the few trees on which we would actually need to undertake full counts ahead of machine harvesting of the remaining population.

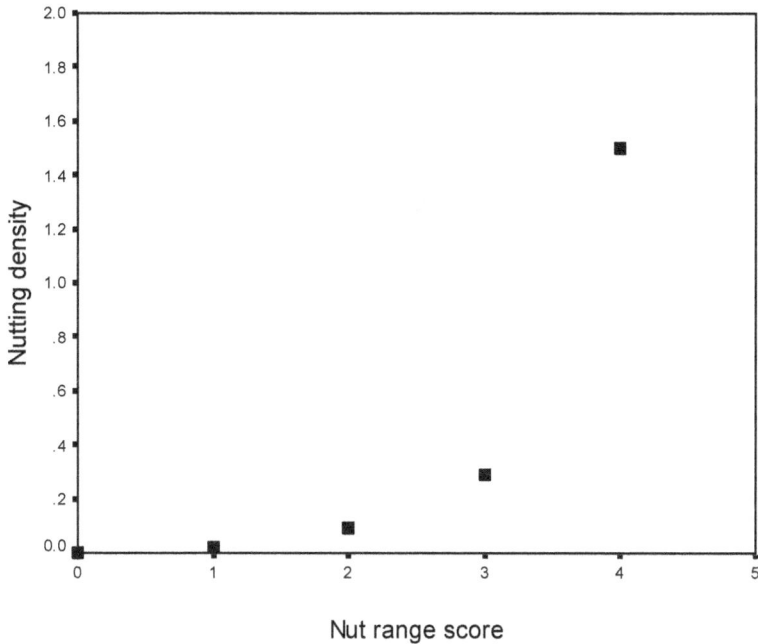

Figure 4-4. Nutting density (NgD) at each NRS, Field 3, Lostwithiel Farm, 2006

Since this original work, conducted in 2006, we undertook a major modification to NRS in 2008, when it was obvious that nut numbers per tree would far outstrip our original ranges . Rather than basing it on a number criterion we have switched to a relative-density criterion: no nuts = 0, low density in the canopy = 1, medium density in the canopy = 2, high density in the canopy = 3. Again NRS were scored by volunteer Ontario Stewardship Rangers and trees were flagged according to the density group to which they belonged. We also switched from NgD to NY as our selection criterion of interest. Nut yield (NY) is calculated as number of nuts per sq. cm of cross-sectional area at DBH. NY represents a move away from our past focus on nutting density (NgD) a variable derived from expressing total nut number per tree against the tree's dynamic height (total tree height less length of clear bole) because of increasing inaccuracy in and time required for measurement of tree height as the tree grows taller. All of our emphasis is on exploring management options for extensive plantations – this means that variables which do not fit with the constraints that extensive plantations present do not remain on the priority list. Calculating NY in this way allows us to compare trees of different sizes (they are all different); DBH is now the sole growth variable recorded annually.

Figure 4-5 shows the numbers of trees per field falling into the different range score categories in 2008. Sixty percent of all trees in the two fields were unproductive (i.e. nut range score = 0) and are not shown. There are successively fewer trees in the higher nut range categories. Thirty-seven trees in total fell in the top category. Within each field the total number of nuts per tree fluctuates considerably across ranges (especially in ranges 2 and 3), but representing them in terms of NY, which takes the size of tree into account, smoothed the data considerably. All average NY_{2008} values are slightly higher for Field 1 than Field 3 (expressed as *100 in Figure 4.5, in order to be able to use a common axis), probably reflecting the one year age difference between the two fields.

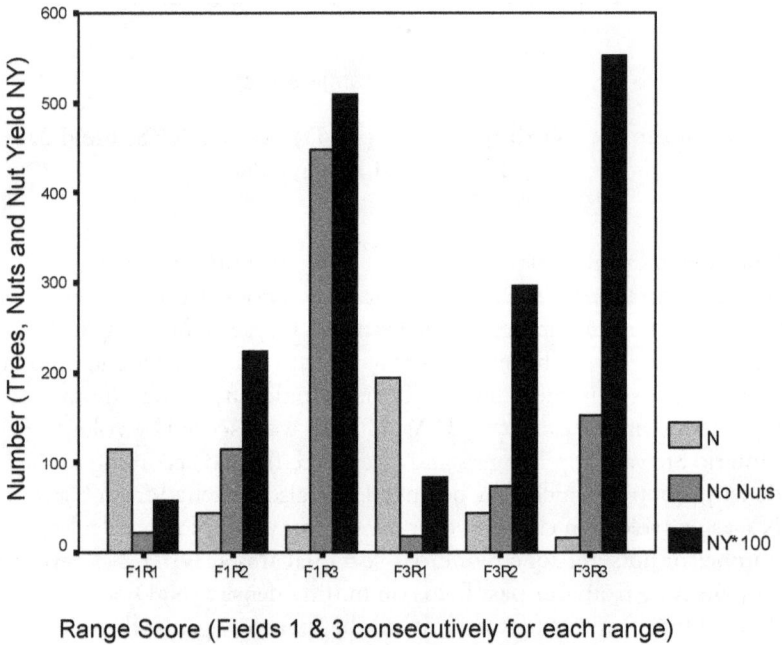

Range Score (Fields 1 & 3 consecutively for each range)

Figure 4-5. Number criteria against NRS for Fields 1 and 3, Lostwithiel Farm, 2008

Kernel yield per tree (KY)

The analysis of 2007 kernel yield per tree (KY, by multiplying KW by the number of nuts per tree; no. of nut data is lacking for two lines), we once again find a re-ordering of lines (Figure 4-6). Line 1-1-14 yielded 1.34kg of kernel. The tree with highest KW (1-1-17) yielded 0.45kg, and the tree with highest K% (1-9-23) yielded 0.32kg. It, of course, depends on our extraction efficiency, but economic yield will be more closely related to KY, and the other parameters of KW and K% become less critical in any selection focus. Because any tree is likely to produce more nuts as it grows bigger, we have to account for this. We will discuss this in the next chapter.

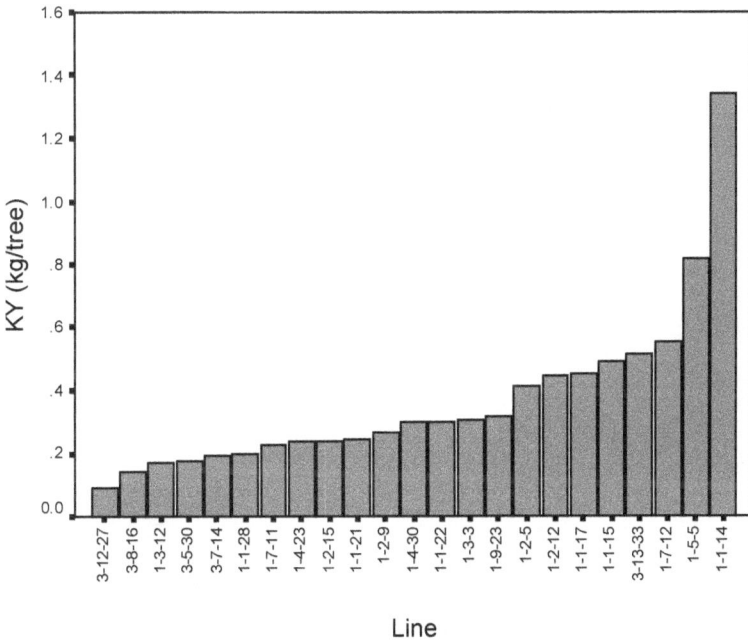

Figure 4-6. Individual KY in measured lines, 2007.

[1] I do not discuss harvesting methods to any extent in this book. We are still exploring options, and it seems to me better to leave this aspect until a later date. Nevertheless, I doubt that we will use a shaker to drop all nuts at once, which we would have to do if we were to count on the ground before mechanical pick-up.

[2] In a recent article in the NNGA AGM meeting in Ottawa, 2007, on the same topic I mistakenly labeled data from Field 1 as from Field 5. I apologise for any confusion this has caused.

[2] They are not truly effects; rather we need to be able to compare NyD at similar values of DBH in order to identify true precocity and fecundity.

5. Gross Nut Yield per Tree

A function of nut weight and number of nuts per tree

Nutting in the Early Years

Probably the most difficult prediction to make is the pattern of onset and subsequent development of nut production. The data used in Chapter 2 to demonstrate potential productivity is derived from an equation that is, quite frankly, useless for young trees, and I have my doubts about its overall relevance to our conditions. However, until we have better we must work with what we have.

How many years will pass before trees begin to produce nuts? This will depend very much on early growth, especially in relation to weed control measures. If weeds are not properly controlled, small black walnut trees can sit in the ground without evident development, sending out small new branches annually but not apparently bulking up individual biomass. Figure 5-1 is the classic photo taken by Fred Van Althen back in the 1980s.

Figure 5-1. Contrast in black walnut growth with and without weed control (courtesy of Ontario Ministry of Natural Resources).

Climate is the other major factor - I have seen five-year old trees in the US bigger than our 10-year-olds, and with more nuts on them. I am not discouraged by this - it is a reflection of our reality - but at times the waiting has seemed endless

Under our conditions, eight to fifteen years appears to be the answer. The range is a consequence of genetic variability in the onset of fruiting, and the fact that every tree, even within our farm, has slightly different growing conditions (imposed by soil variation including drainage, and local weed effects). Furthermore, annual pruning will also influence growth, and as every tree is pruned according to its needs, this is a further source of variability.

Another factor is supremely important. Young trees are extremely susceptible to spring frosts, which will nip off developing buds and flowers. As the trees grow, I think they are marginally less susceptible because of the temperature inversion with increasing height. A few days variability between trees in date of leafing (and, by correlation, flowering) will make a big difference in years to first nut, and subsequent nut yield. Our sentinel site began to produce nuts (16 trees out of 590; ten years after planting) in 2004. In May 2005, just as trees came into leaf, and catkins were appearing everywhere, we had two successive nights of nearly -5°C. Nothing was left on any tree already leafing out, and they all had to regrow from secondary buds. Fruiting, however, was completely curtailed.

A spring frost, occurring at bud emergence, can play havoc with the rest of the year. I have noticed that at Lostwithiel Farm the critical period is the first week of May, and that if we can get through it with nothing less than 2°C our course should set fair. What is the likelihood of this happening? In Figure 5-2 I have graphed both date and temperature of the latest event in May of equal to or less than 2°C over the nine-year period for which data is available at a station to the east of the farm.

What we immediately see is a high probability of a damaging event occurring at least mid-month or later, and that we should not be surprised by any late spring frost. The 2002 event, a negative reading on the May 20th, would have tested anyone's resolve. The 2005 year, as noted above, was similar. While Figure 5-2 doesn't show it, in 2009 there was a double bottom, with coldest (shown) recorded on the 11th, but another cold wave (with a 2.6°C minimum) recorded on the 26th. It is likely that the latter finished off whatever survived the former, and hence there were no nuts in 2009.

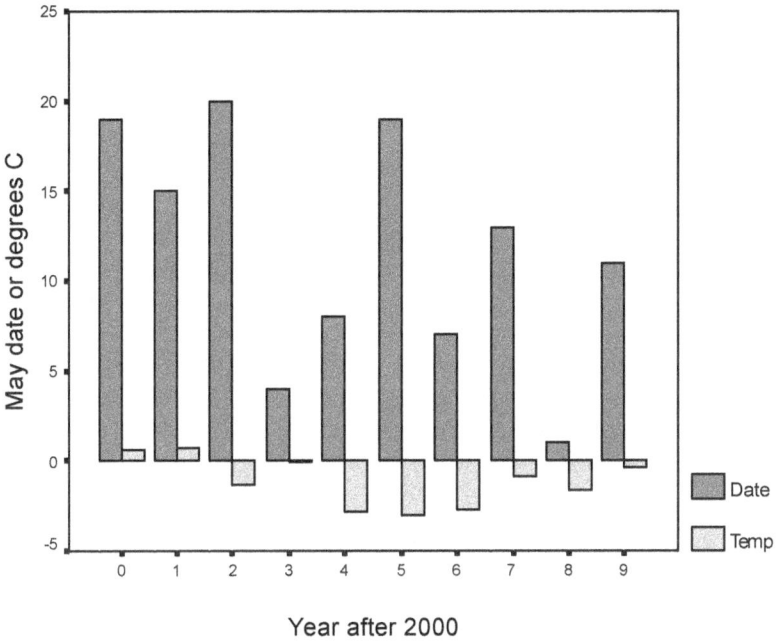

Figure 5-2. Date and minimum temperatures at 2°C or less, May 2000-09, Kemptville Ontario.

Every year is different in the way spring growth is stimulated, and because of this the trees present different degrees of predisposition to frost damage. Figure 5-3 shows the accumulation of growing degree days (above 10°C) for the years 2001-2009 at the same weather station as Figure 5-2. These two figures need to be interpreted in parallel. For instance Figure 5-3 shows 2008 to have got off to a warm start; Figure 5-2 shows that the last truly cold event of the same year was early in the first week of May, before flowering buds could be severely harmed. By contrast, 2005 was more or less average in growing-degree day accumulation, but had a cold event around the third week of May – fruiting was poor in 2005. 2009 showed a below average start, with the first of two May cold events in the second week. The result has already been noted.

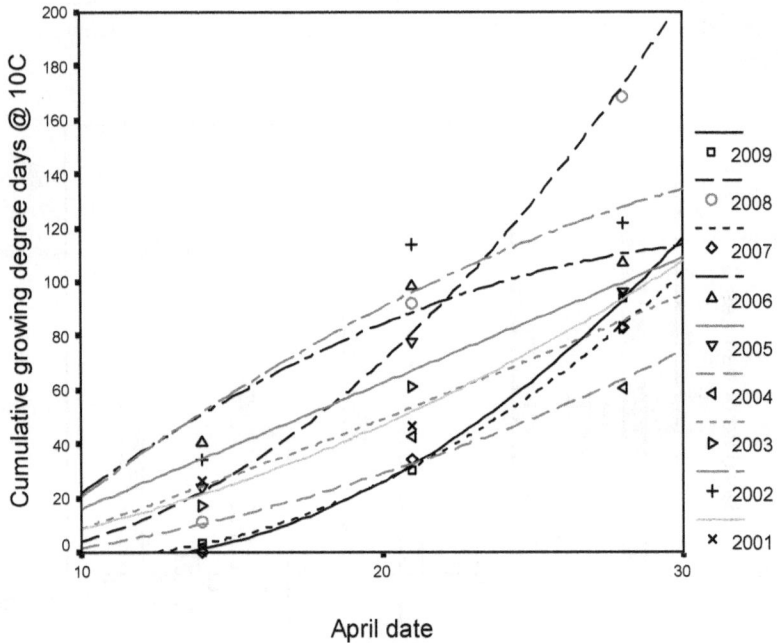

Figure 5-3. Cumulative growing degree days (10°C) April 2001-09, Kemptville Ontario.

However, another factor to consider here is tree size. I collected nuts widely throughout our region in 2002, in fact, it seemed to be a bumper crop, for, before 2008, I had not seen the like since. Frosts generally occur as a result of a temperature inversion, and my hypothesis is that a tall tree has most of its canopy above the freezing zone. What I hope eventually to see, as our trees grow taller, is a declining effect of spring frosts on our own nut production. It is possible that, when young, many of the trees out in the landscape have been naturally selected for late flowering, though I don't place much confidence in this. As far as I know, our first-planted trees are derived from nuts collected in our own seed zone. I actually believe lateness in bud emergence (and thus flowering) runs counter to the natural trend, and that earliness is of evolutionary advantage, but this is nothing we shall prove at the farm. What we will test is the heritability of date of bud break in the 25 lines selected in 2007, and see whether there is any consistent lateness between maternal and F1 lines.

In the previous chapter I mentioned that about 60% of our trees were still non-producing. What we don't, of course, know is the percentage of our stands which is too early for our environment i.e. trees that are physiologically mature but produce no nuts by reason of late frost effects. We focus on the actual bearers, which may be relatively late in flowering. It would take a dedicated couple of weeks at least to watch all trees and score them on flowering, and this we haven't done.

Such effects are an important component of what is described as the Alternate Bearing Index (ABI). The ABI describes the year-to-year variation in nut yield of a tree, which also has its genetic component, i.e. even under perfect conditions the tree would probably not yield consistently year-to-year (this is difficult to prove, especially under the conditions of increasing climatic variability, as we appear to be facing now). Nevertheless, the ABI will be one of the biggest factors influencing productivity. The experience of 2005 and 2009 tells us that there may be annual macro-swings in productivity (every tree affected), as well as micro-swings (variation between years within a single tree when there are no macro effects).

The average number of nuts per tree produced in 2004 was four (the range across the 16 producing trees in Field 3 was from one to 25). Four nuts is perhaps 100 g; 25 is still likely to be less than 1 kg. Obviously, we will not make any money in that first year from nut production. It will be a time of curiosity (it was for me) with, perhaps, some conviction that there will be more nuts the following year. I was completely wrong (actually not quite, because there was one tree in the sentinel site which gave me one nut in 2005), frost struck, and I had to find the patience to wait until 2006 to measure progress. One hundred and fifty trees in Field 3 nutted, at an average of just over 13 nuts per tree; mean field DBH was 4.4 cm, with nutting trees averaging 5.7 cm. By contrast, 112 trees of Field 1 nutted; population average DBH was 4.9 cm; nutting trees averaged 7.7 cm giving 50 nuts per tree.

However, it was 2008 when things began to turn around. About half the trees in Field 3 nutted, with an average of 35 nuts per producing tree. Average DBH across the whole field was 6.6 cm, while nutting trees had an average DBH of 7.5 cm. Again, by contrast, 230 trees in Field 1 nutted (an average of 100 nuts per tree), with an average DBH of 13.7 cm compared to a populational mean of 6.6 cm. All these values are expressed in Figure 5-4.

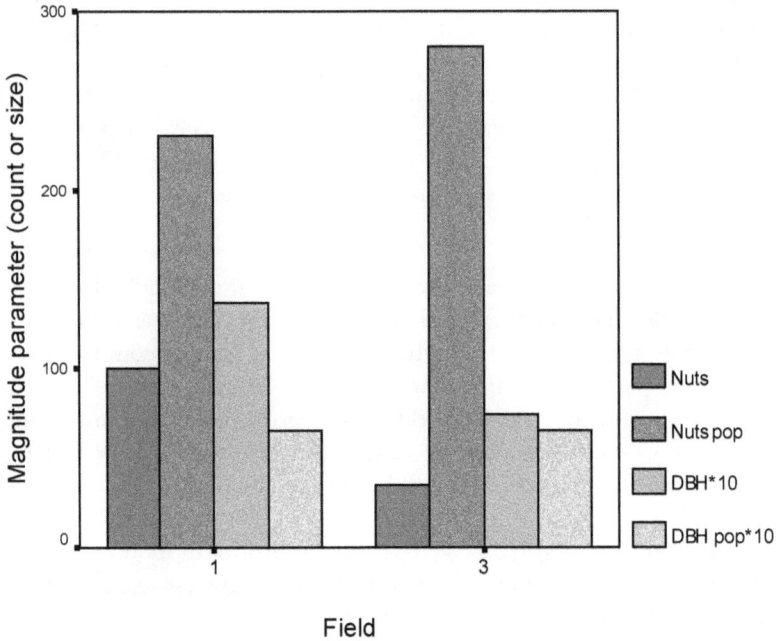

**Figure 5-4. Nutting milestones, Fields 1 and 3,
Lostwithiel Farm, 2008.**

From these data we can say that if the season is appropriate for flowering
bud survival, at our site, trees at or above a mean DBH of 5.5 cm should
produce fruit, with approximate nut numbers on average increasing by
about 20-25 nuts per cm of DBH above this, for the range observed. I
add this caveat, because each tree will be different; because above this
range nut counts are much different, and I identify elsewhere the
complications associated with using DBH as a comparative measure even
though I derive another measure from it.

I actually need to go back a little, and discuss the probable differences
between grafted and seedling trees. Because a grafted tree is produced
with mature wood, it will begin fruiting almost immediately. A small
grafted tree will not, however, produce many fruiting buds, and overall
nut production in the first few years will not be economically significant
because the tree will not have a big enough canopy to sustain heavy
nutting. As we do not have any grafted black walnut on our farm we have

no data which is directly comparable; however, we can look at some of the experiences of others and answer the questions we may have.

Before I do so this let me introduce the term heterodichogamy, which, while it sounds complicated is no more than the term describing the relative time of emergence of the male and female flowers (yes, they're separate) on the black walnut. When I talk about flowering bud survival for nut production I am really focusing on the female flowers for it is here where the nuts arise, and so is what I am relating to DBH. Somewhere, on some tree, catkins will be releasing pollen into the plantation, for fertilization of the female flowers on other trees (a defence against in-breeding).

But on to others' experience. In one study which examined nut production in 21 grafted selections in a warmer part of the US over the age range 10-14 years, average nut production was 4.3 kg/tree[1]. It was noted that nut production began four years after propagation, but that the first significant crop was not obtained until the tenth year. Interestingly, there was clear evidence of the ABI, with average yields in successive years being 5.4, 1.8, 5.7, 2.2 and 6.3 kg nuts/tree. Alternate bearing was absolute in some selections, and weak in others. The trees were fertilized annually, and sprayed to control anthracnose, one of the principal diseases of black walnut.

Another study[2] looked at the difference between average nut yield per tree of a biomass stand, and the nut yield of the highest producer within that stand. A further variable, the quality of the growing site (high vs intermediate), was included, to highlight another difference. Average yield at ages of 14 (high site) and 15 (intermediate site) were 1.3 and 0.8 kg nuts/tree, respectively; the select trees at each site produced 11.7 and 4.6 kg/tree, respectively. The authors of this study make some simple conclusions about the effects of site quality, and what it would mean if the whole plantation were made up of the select trees encountered (this could be done through grafting). In economic terms they suggest a select plantation would return nine times per hectare what a non-select (i.e. biomass) plantation would return.

Firstly it is important to note that a single year's data is an extremely unwise basis for drawing conclusions. Previous mention drew attention to the importance of the ABI in interpreting nut yield; the latter study does not address ABI and the possibility of genetic differences between the stands accounting for what appeared to be site differences. Secondly, the concept of a select tree as the benchmark for site potential is fraught with danger. Perhaps select trees presented absolute ABI but were sampled in the 'on' year; perhaps there were microsites within each plantation which were even more favourable than the stand average. The concept would have had more validity if it were clear that the select values were the average of the five or ten best trees, rather than of a single individual.

A little can be concluded from these studies (but remember that they were conducted under US conditions): a grafted plantation might yield an average of 4.3 kg/tree by 15 years of age, while our common biomass plantation might only yield 1 kg/tree. We would then still want to ask the question whether a 15 year-old tree matches the 15cm DBH tree of interest to Zarger (1946; yielding about 1 kg/tree), and whether subsequent production will follow the trend he suggested. It will be the answer to the latter which tells us whether we could forego the significantly higher expense of a grafted plantation, and work with extensive plantations of cheaper non-select trees.

In the previous chapter, we saw that our top line yielded in excess of 1 kg of kernel in 2007, when it was about 15 years old. In 2008 a different line yielded in excess of 1,700 nuts (I know because I counted them all). In 2007, this tree recorded a nut weight of 15 g with a K% of 26%. Because I could not measure kernel production in 2008, extrapolating from the 2007 data suggests that 1700 air-dried nuts and kernel would, in 2008, have weighed 25.5 kg and 6.63 kg, respectively. The tree had a DBH of 16.7 cm at the beginning of the 2008 season, close to what Jones et al (1995) discuss for their select trees. It was 16 years old. When it was 15 years of age, it yielded in excess of 500 nuts, or 7.5 kg and 1.95 kg of air-dried nut and kernel, respectively.

Correcting Nut Yield for Tree Size

Every tree in a walnut stand (orchard or plantation, grafted or not, it doesn't matter which) is different in terms of growth. Some trees grow faster than others. Some give more nuts than others. Most of this growth difference in is measurable as annual DBH change, which allows us an estimate of absolute growth and growth rate.

DBH measurements offer the simplest way to adjust number of nuts per tree for this biological difference in size. Before, while we were still recording height, we calculated a nutting density, NgD, which was an expression of the number of nuts per cm of canopy height, itself calculated by subtracting clear bole length from tree height. But this required two extra measurements, height and clear bole length. As we had found a clear relationship between D and DFT, measurement of DBH serves as a proxy for H, and it is far simpler to calculate an index of number of nuts per unit of cross sectional area (CSA) at DBH, or per cm of perimeter at DBH. We use the former and call it nut yield, NY. Obviously as DBH increases, CSA increases exponentially at a rate π times greater than perimeter, thus offering greater sensitivity than a perimetrically-based index. But it doesn't really matter which you use as long as you are consistent and use a common method for all trees.

Then comes the debate on whether it should be number of nuts or nut weight that we measure. As we collect data on number of nuts, and use of total nut weight would require that we eliminate the hull component, number of nuts is far easier, though a weight-based index might replace it over time and as cleaning processes become simpler. For instance, I can imagine using a weight-based index to make ultimate decisions about tree selection. While we have shown that, in 2007, our '3' trees had an individual NW range of about 7.5-15 g (Figure 3-7), this was calculated after hulling, washing and drying samples, and the question is whether this has any effect on our calculations of number-based NY.

The number-based nut yield per tree (NY) for the group of superior lines for the period 2006-2008 (n=51) is shown in Figure 5-5. Over this period and in this group, average NY increased three-fold while average DBH increased by about a third.

Finally we come to the question of superiority, or useful line (as raised in Chapter 4). Table 5-1 answers the question 'How many trees appeared more than once in the top five trees by parameter?' While five lines appeared more than once, only 1-2-12 appeared in the top five in almost all categories. I'd suggest that all these trees are candidates for superiority, but that only one is truly outstanding (I will add the caveat 'as measured in 2007', because 2007 may have been an 'off' year). For reasons I have already explained, repeat measurements of kernel parameters could not be made in 2008.

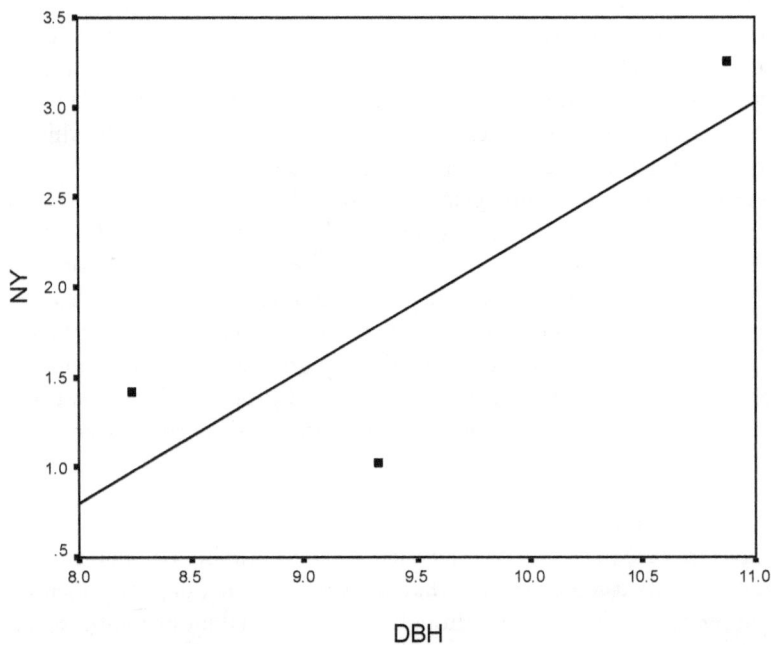

Figure 5-5. NY plotted against DBH for superior trees, Lostwithiel Farm, 2006-2008.

Parameters	NW	KY	K%	NY
NW		1-1-14, 1-1-15	1-2-12, 1-3-12	1-2-12
KY				1-7-12
K%				1-2-12
NY				

Table 5-1. Superior lines by parameter, Lostwithiel Farm, 2007

The Need to Correct Alternate Bearing Index

As mentioned above, black walnut is known for its tendency to produce more heavily on a biennial cycle. This is commonly referred to as alternate bearing, for which an index (ABI) exists. Calculated as a value between 0-1, the ABI is a measure of the consistency in nut production, but is subject to misinterpretation for two reasons - each requires an adjustment factor.

Adjustment 1 (for NY)

The first adjustment required is for DBH, so accounting for tree size, and calculate the index from NY, thus more truly ABNY, to account for change in potential productivity across time based on tree growth.

Adjustment 2 (for regional environmental events)

The second adjustment is for broader environmental influences which affect the population as a whole. The Hammons Crop Update for 2006[3] notes:

"The 2006 Black Walnut harvest was successful. Buying at our 260 hulling locations in 16 states has ended….. The total volume is above average at over 28 million pounds. This is down from the 2005 harvest of 36 million, but well over most "short crop" years in the black walnut's alternate bearing cycle….."

Where did most of the nuts come from that year? The heaviest areas were southern Missouri and northern Arkansas. Other regions were not nearly as busy. Central Missouri and most other regions produced a much shorter crop than the previous year. Kentucky and Tennessee were fairly light. Hammons expected that 2007 would be good for those areas.

It is surprising that this fluctuation apparently requires no elucidation. There is no attempt to explain why Missouri and Arkansas were heavier, or Kentucky and Tennessee were lighter. The underlying rationale, though unstated, seems to suggest that the built-in bienniality of black walnut is to blame, and that no questions need to be asked. But we can fairly confidently infer that alternate bearing must have a large environmental component if it expresses itself widely at the multi-state level.

If we calculate an ABI for a tree irrespective of the population's performance, we are isolating a series of events from some of the reason for its expression. Comparing ABI values between trees without reference to the seasonal effects regionally is entirely misleading. In our case it is complicated by the fact that the trees are reaching physiological maturity and thus increasing numbers of them 'should' be producing in successive years. If they don't, we need to separate between delayed onset of maturity and environmental effects on bearing.

The wider environmental effect was noticed at the farm in 2009, where there were no nuts on the any trees after a harvest of 1.5 t in 2008. The reason for this has been dealt earlier within this chapter. But I collected from a site in southern Ontario where a fair proportion of the 1600 trees at the site were bearing. It is inappropriate to assign the effect of the very late cold event at the farm in 2009 to a performance factor related to the genotype itself[4]. 'Zero' performance will markedly affect a calculated ABI value for any tree, and distort a 'true' ABI (which is an annual variability in fruiting expression not compounded by the environment).

The nature of the ABI calculation (remember, we use ABNY values) is:

$$I = 1 / (n\text{-}1) \times \{\,|\,(y2 - y1)\,| \,/\, (y2 + y1) + |\,(y3 - y2)\,| \,/\, (y3 + y2) \ldots + |\,(y(n) - y(n\text{-}1)\,)\,| \,/\, (y(n) + y(n\text{-}1))\}$$

where n = number of years, and y1, y2, …, y(n-1), yn = (nut) yield of corresponding years; the vertical bar in some terms implies an 'absolute' result where the sign is always positive.

Ideally, we should introduce a term for the population mean for all trees and calculate a further adjustment for ABNY, where our individual result is relative to the population as a whole. Then, of course, comes the question of how far the population extends. The ABNY for the population of '3' trees was 0.23 for the period 2006-2008; we'll call this $ABNY_{pop3}$. This can be interpreted as saying that on average our '3' trees bear less on a biennial basis than they do on an annual one; but this value will be forced much higher towards one if we include 2009, because no trees bore nuts in 2009. We need to express individual ABNY relative to the population as a whole, i.e. $ABNY_x / ABNY_{pop3}$, and we'll call it a relative ABNY, for an individual x in the measured population of '3' trees. Obviously, the period should be common. Is this as intuitive as the 0-1 range of the original ABI? I don't yet know, and I haven't enough data yet to know if this is the solution. In Figure 5-6 I have graphed the relationship between these two variables for the five lines shown in Table 11-1, for both periods 2006-08 and 2006-09, and the straight lines gives me some confidence, though they also imply that all we have done is change the scale on which we interpret the original ABNY. Obviously the more severe was the period (or a year within it), the different the slope. But this is also probably affected by the number of years covered. It will take multiple site comparisons to resolve the interpretation.

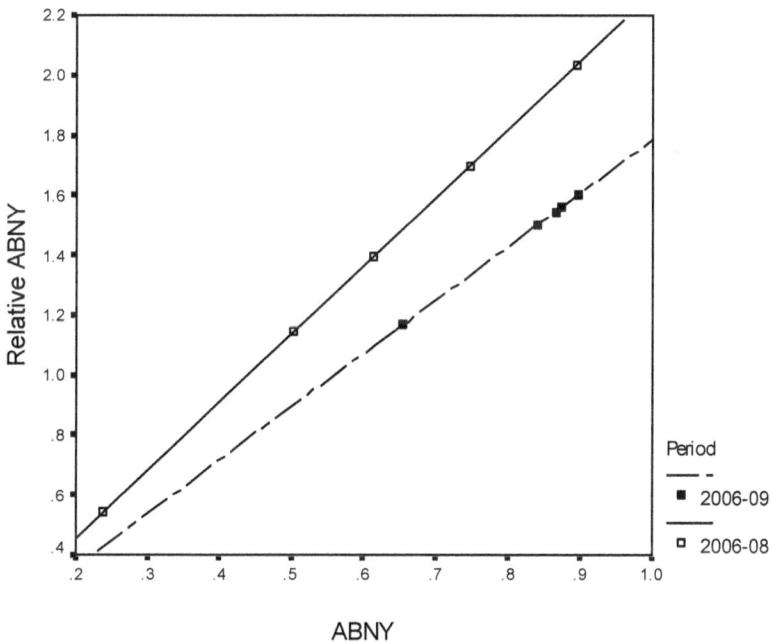

Figure 5-6. Relative ABNY of five superior lines, Lostwithiel Farm 2006-2009.

But I am convinced that both of these adjustments are required if we are not to misinterpret ABI.

[1] Reid, W., M.V. Coggeshall and K.L. Hunt, 2004. Cultivar evaluation and development for black walnut orchards. In: Michler, C.H., P.M. Pijut, J. Van Sambeek, M.V. Coggeshall, J. Seifert, K. Woeste, and R. Overton, eds. Black walnut in a new century. Proceedings of the 6th Walnut Council research symposium, July 25-28, 2004, Lafayette, Indiana. In Gen Tech. Rep. NC-243, St Paul, MN. US Department of Agriculture, Forest Service, North Central Research Station, (188p.) 18-24.

[2] Jones, J.E., H. E. Garrett, J. Haines and E.F. Loewenstein, 1995. Genetic selection and fertilization provide increased nut production under walnut-agroforestry management. Agroforestry Systems, **29**: 265-273.

[3] http://www.black-walnuts.com/page.asp?category=information&subcategory=cropupdate

[4] This was a response to flowering date rather than anything else.

6. Nut Yield per Hectare

A function of nut yield per tree and number of trees per hectare

I do not want to say a lot about number of trees per hectare as one of our determinants of nut yield per hectare, except to say that the decision on density has a major effect on the later situation of where some trees have to be removed (i.e. thinned out) to give the others more room.

But let's start off with the issue of planting density. Because early nut yield per hectare will be driven by the number of trees per unit area and the productivity of those trees, we want to maximize density in the early years to make up for the low productivity of young trees. We have worked with two planting densities: 20' x 40' and 20' x 20'. This translates metrically to 6 m x 12 m and 6 m x 6 m. The lower density was used in Field 1, but I soon went to the higher density when it became evident that my neighbour had a finite limit for his hay needs, and that only one field needed to be at a density where large machinery could easily work.

If intercropping is something you intend to practice, then judge row width on the basis of some multiple of typical machinery size, plus a foot or two for the inaccessible spots right under the rows of trees in the early years. I refrain from putting a number on this, because everyone will have a number they are comfortable with, and this should be the basis for decision. In my other fields I can run my 185cm disc mower up and down the rows, including under the tree canopy, and three passes gets the grass mowed. It keeps me sufficiently far out to avoid clipping tree trunks. Before harvest I repeat the operation, both lengthways and crossways, so that the nuts are falling into short grass. The crossways pass cuts those islands or files of tall weeds that were not cut in the summer.

The other issue with machinery is that of pruning height. My rule is to leave as many branches as I can on young trees, using an herbicide for vegetation control. Don't plan on getting machinery under the canopy until the trees have a decent canopy to mow under, or you'll end up mowing down the trees! It can be hard to discern lateral distances from a tractor seat. My pruning management is such that the average annual increment in clear bole length is something like 10 cm calculated across a whole field. Not conducive in the early years to mechanization of tasks directly around the trees.

Needless to say, adequate headlands should be provided in order to turn even a small tractor. This will be at least 6 m, and for large machinery almost certainly double this. Tree spacing within the rows is not as critical a factor.

While I don't think tree spacing within the row is such a critical factor, the mature walnut tree is said to require a 40' (12 m) diameter area, so using planting spacings which are simple fractions of this will allow you to thin to that 'ideal' mature density. If you plant at 3m spacing (between and within rows), however, you will have four times as many trees per unit land area than if you plant at a 6m spacing, and 16 times as many trees than at a 12m spacing. The closer spacings will offer you considerable scope later on for selection of the better nut producers. If you plant at a closer spacing than 3m, you are approaching a forest system, and the stand may reach full closure before nut production occurs. Heavy thinning would then be required to restore significant nut production.

Thinking about a thinning strategy

One of the working principles of a biomass nut plantation is that a significant proportion of the trees originally planted will have to be removed in order to allow the remainder to continue to expand. If thinning is not undertaken, the plantation will eventually metamorphose into a timber stand as the trees grow rapidly in height, and nut production declines.

The challenge in the thinning strategy is to identify the parameters that will help one make an objective decision. In fact, in a biomass strategy, one has to make thousands of decisions because of the relative importance of each tree to outcome[1]. At Lostwithiel Farm we are not yet at the point where any thinning is necessary, but this is such a fundamental operation that it is important to understand how to make the decisions, and to be sure that they are properly based on on-site data. What I describe in this chapter is a synthesis of ideas, and I hope through analysis to arrive at something which, at this moment, looks like a good bet.

Our thinking about a thinning strategy has encompassed the following:

- Recognizing the probable importance of a link between tree growth and productivity – a tree that is growing well is likely to be more productive than one that isn't, because of the rate of canopy expansion.

- Recognizing other factors which might detract from productivity, such as relative incidence of leaf-spotting diseases.

However, there are also other issues probably closely linked to individual genetic characteristics which may be important, such as nutting precocity[2]. The issue for the walnut grower is how to include these factors in some sort of decision mechanism, so that thinning leaves the best trees and doesn't bring about their removal.

One of the aspects that must be remembered is that every tree is growing in relation to its neighbours. Consider Figure 6-1. In a square layout, with neighbours equidistant, there are four close neighbours to every tree (inside each cross, e.g. nos. 0, 1, 2 and 3), and four less close (the diagonals just outside the cross). The exceptions are corners and edges, where there are two and three close neighbours respectively (nos. 5 and 4). These neighbours, by their own growth, influence the growth of the tree at the 'centre' of the cross[3]. But in Figure 6-1, every tree at least one row in from the edge is at the centre of a cross, not just the ones indicated.

Other planting patterns might require different treatment. The most obvious is that where there is much greater distance between rows than between trees within the row. This would be probable under an agroforestry system where crops are grown between the rows. In such a case we only need to think about the two (within-row) immediate neighbours on either side of each tree, or at least until between-row interactions become important in later decades.

How then can we calculate an index which separates trees, so that we know which ones to keep and which to thin out?

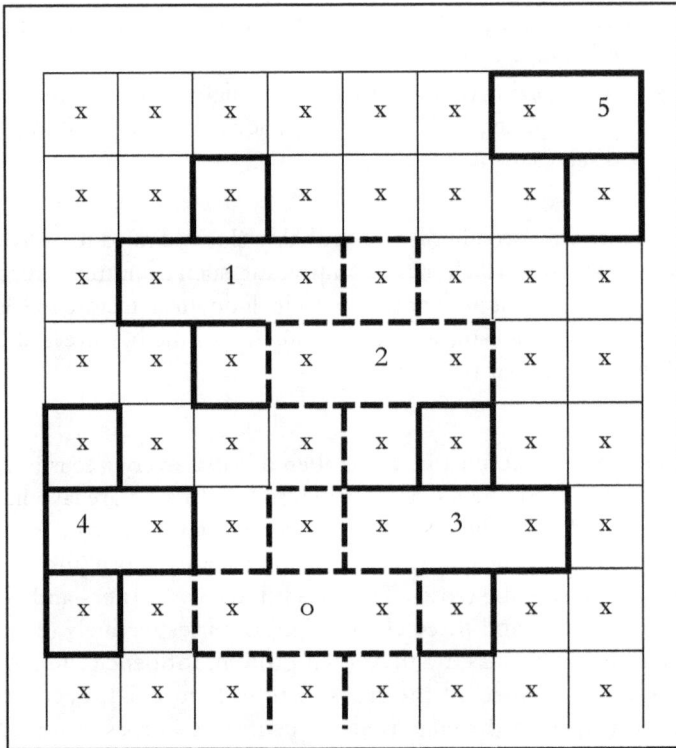

Figure 6-1. Neighbour relationships in a square planting layout.

In very general terms we are going to have to decide which variables or single variable best express(es) the productivity of the trees in our stand. If we do not need to thin until all trees have reached onset of nutting, it may be quite easy to identify our best trees, but considerably more difficult to decide where the thinning stops. The corollary of being able to identify our best trees is that we are also able to identify our worst trees. However, there will be a range of trees that spans the best-worst continuum, and the importance of our strategy more truly relates to how we will thin this continuum, assuming that all the best trees stay, and all the worst go.

I'm going to illustrate this using a single variable, though a host of arguments have run through my mind on occasions for the different variables that we could include, and you may say the same. In the end, though, we need to focus on the variable that is the best expression of nutting, without the complication of the multitude of variables which could be said to influence nutting, because we would have the extra complication of deciding how the factors should be weighted amongst themselves to accurately express their relative influence on nutting.

Let's start with the single characteristic which we will have been measuring annually, the NY. Just as we cautioned about calculating ABI on too few years data, our NY would be better calculated as a multiple-year average. Let's call this the MNY, and please assume from this point that this is what we use. What we then want to know is: what was the MNY of each tree relative to the MNY of its close neighbours? The following example actually uses the NY data of a group of trees in Field 1 for the period 2006-2008 (Figure 6-2). Also shown in Figure 6-2 for each year is the NY_{pop} and DBH for all trees nutting (n).

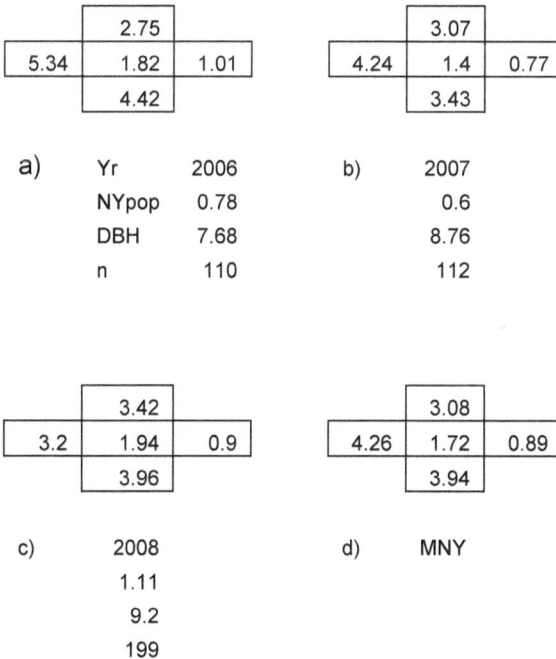

	2.75	
5.34	1.82	1.01
	4.42	

a)

Yr	2006
NYpop	0.78
DBH	7.68
n	110

	3.07	
4.24	1.4	0.77
	3.43	

b)

2007
0.6
8.76
112

	3.42	
3.2	1.94	0.9
	3.96	

c)

2008
1.11
9.2
199

	3.08	
4.26	1.72	0.89
	3.94	

d) MNY

Figure 6-2. Neighbour NY relationships for calculating MNY, Lostwithiel Farm, 2006-2008

For every tree we calculate the MNY (d, Figure 6-2 and a, Figure 6-3). But then we need to go further, because we want to know the MNY of the neighbours as a whole, rather than as four individual trees, so that we can interpret a competition factor between our trees. So then let's take the average of those four values (3.04, b, Figure 6-3; call it neighbouring NY or NNY) and compare it to the MNY of the centre tree (1.72, d, Figure 6-2). As the latter is lower we can say that our centre tree is less productive for its size than the average of its neighbours, but if it were higher we would say that our centre tree was relatively more productive for its size. In fact, the easiest way to represent this is by substituting this comparative value (CNY) for the original MNYs used to calculate it (-1.32, c, Figure 6-3). All this is very easy to do with spreadsheets.

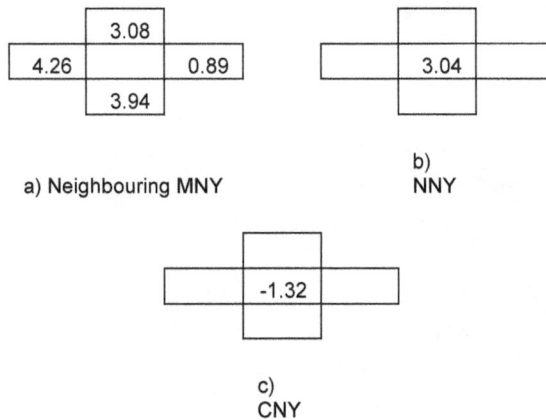

	3.08	
4.26		0.89
	3.94	

a) Neighbouring MNY

	3.04	

b)
NNY

	-1.32	

c)
CNY

Figure 6-3. MNY and CNY example, Lostwithiel Farm

What are we going to do with the CNY? It is our base factor for selection, and across our field we want to know the CNY range so that we can see which were relatively the most productive taking into account their size (as NY does). Figure 6-4 indicates the actual values for the surrounding trees in Field 1 including the example we just worked through.

The range in CNY values in Figure 6-4 is from 4.99 to -2.57. This a total range of 7.56. If we divide this into four equal sub-ranges, we can categorize our rankings as follows: 4.99 to 3.10, high, 3.09 to 1.21, medium, 1.20 to – 0.68, low, and -0.69 to -2.57, very low. We could then say categorically that all high trees will stay, that all medium trees will probably stay, that low trees are possible candidates for thinning, and that very low trees are our probable candidates for thinning.

1.60	1.67	0.28	0.01	0.56
0.10	3.46	0.07	0.01	0.03
0.17	0.73	1.69	0.13	0.53
0.21	1.94	0.01	5.25	0.11
0.69	3.08	0.24	0.81	0.01
4.26	1.72	0.89	0.70	0.01
1.10	3.94	0.01	0.01	0.01
0.01	3.96	5.49	0.60	0.01
0.95	1.16	0.79	0.12	0.01
0.87	0.04	3.32	0.22	0.01
1.75	2.05	0.01	0.46	0.05

a) Calculated MNY for all trees

0.64	1.36	0.06
1.82	0.23	1.87
1.01	2.28	0.27
1.15	1.20	1.55
3.04	0.67	0.43
1.70	2.58	0.33
2.65	1.34	1.41
1.44	2.52	0.40
1.85	0.27	0.98

b) Calculated NNY values

2.82	-1.29	-0.05
-1.09	1.45	-1.74
0.93	-2.27	4.99
1.93	-0.96	-0.74
-1.32	0.23	0.27
2.24	-2.57	-0.32
1.31	4.15	-0.81
-0.27	-1.73	-0.28
-1.81	3.05	-0.76

c) Calculated CNY values

2.82	-1.29	-0.05
-1.09	1.45	-1.74
0.93	-2.27	4.99
1.93	-0.96	-0.74
-1.32	0.23	0.27
2.24	-2.57	-0.32
1.31	4.15	-0.81
-0.27	-1.73	-0.28
-1.81	3.05	-0.76

d) Resulting thinning
pattern

Figure 6-4. Example thinning pattern, Field 1, Lostwithiel Farm.

Of course, this is based on projecting past performance forwards. I think there is a strong argument for doing this and I am not going to suggest there's an alternative. NY is an outcome variable, i.e. it integrates all the effects of phenology of flowering, disease impacts, site effects, tree size and possibly others on our main determinant of productivity in a developing and developed stand. I should say that some trees are still unproductive. Where you spot MNY values of 0.01, it means that the tree in question produced no nuts in any of the three years of data used in this example. Spreadsheets dislike non-zero values, so entering a value of 0.01 reflects the closest thing to a non-productive tree and does not affect the result.

The thinning result can be found in d, Figure 6-4. Very low CNY trees for immediate thinning are indicated by the dark grey cells; lighter grey cells indicate those trees which fall into the low CNY category, and which could be removed on a second wave of thinning after recalculating CNY with a couple more years of data. The same is true for the medium category (CNY 3.09 to 1.21) if a third wave of thinning is required some years hence. My suspicion is that ABNY will become a bigger factor in evaluating MNY, and so CNY, values of the higher categories.

[1] Thinning strategies in an orchard system, with a few named selections, may be different: retention of selection balance (i.e. removal of equal numbers of each, if the orchard began with such a distribution), or removal of certain selections if experience has shown that certain selections are not performing well. The assumption here is that micro-site effects within selections will be minimal, otherwise one might as well treat the orchard as a biomass plantation.

[2] Precocity is often used as a measure of earliness in flowering or fruiting. I use it as a descriptor for fecundity, or prolificacy in fruiting, and so not necessarily related to earliness. Because earliness may actually be a disadvantage to fruitset on the northern fringes of the species range, I do not want to give the impression that we are selecting for earliness if we select for precocity. I think of a tree as precocious if it starts fruiting at a young age rather than 'early' in the calendar year.

[3] I have already noted that we are not yet at the point where we have to start thinning. At present, growth is likely just the outcome of individual GxExPP, where E may still lack a competition component. However, this does not detract from the need to be able to detect such influences as and when they occur

7. Kernel Percentage

A function of genetic potential and growth conditions

Kernel percentage is the variable that defines the majority of named selections (I am going out on a bit of a limb here, as someone will no doubt have an argument to show that this isn't so) and so drives the horticultural industry's interest in black walnut. Other variables may also be of interest, which is why Kwik Krop is probably called Kwik Krop, but US state-level 'cultivar' evaluations tend to emphasize K% above all else, even before issues related to NW[1].

I have mentioned in Chapter 4 what K% tells us. Not a lot, actually, because the associated ease of cracking and separating ('crackout') has not been found to be linked directly to K%, and this is what we really want to find. Hammons notes[2]:

> *Some improved varieties produce nuts with thinner shells and bigger nutmeats than wild nuts. With this increase in nutmeat to shell ratio, the nuts are worth up to five times as much as the wild crop.*

This implies that they are working with a low average K% in the wild harvest they buy annually (I recall 6% being mentioned anecdotally, which must be largely driven by the nutfill switch I have previously mentioned), but is it true, therefore, that wild (or non-select) types all show low K% and that only the 'cultivars' yield a high enough K% to make black walnut production economically worthwhile?. Let's take a look at the results of the 2007 assay (see Figure 7-1).

The highest recorded K% was over 28% (line 1-9-23). This was far higher than I expected to find in this sample. From my off-farm study I had expected a mean K% of about 20%, with perhaps the top line approaching 24%. But it is actually quite exciting to find 28% kernel, because the probability is that we shall eventually find higher values (my off-farm study begun in 2002 identified more than one tree with more than 30% K). Again, K% is not correlated with NW or KW. In fact, on-farm, the highest K% was found in a line with mean NW <10g. The highest K% line 1-9-23, KW 2.7g, is at far right in the histogram in Figure 7-1.

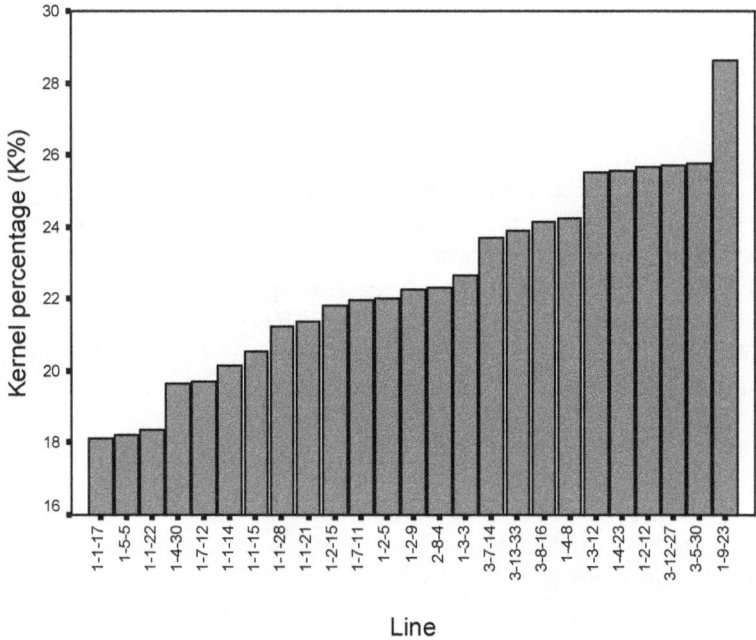

Figure 7-1. Individual K% in measured lines, 2007.

Grafted named selections in the US have yielded K% of up to 39%, though averages tend to be a bit lower than this[3]. Our experience in 2007 indicated the existence on the farm of lines of up to about 75% of this potential K% in common walnut obtained from our own seed zone. We have planted an F1 trial of these 25 lines to examine heritability in K%, and the other traits measured on each tree. Almost all of the 25 lines had good disease range scores (Chapter 8); it will be interesting to observe heritability in this trait[4].

I think I am expressing a truism when I say, that apart from Hammons, the attitude generally in North America is that you can't hope to make money without using named selections. I think I am also right when I say this must be patently untrue. The ultimate K% of the Hammons and orchard models are just too far apart for this to hold water. Obviously Hammons works at a scale no-one can attempt to repeat, so the amount of money one could hope to make is quite a bit less, but I would suggest that cracking and separating efficiencies can go a large way to making up the difference.

The premise underlying our biomass approach to nut production is that there will be enough trees in the overall population with characteristics (traits) of sufficient value to exploit multiple income streams. To date we can characterize these streams as kernel, shell, and sequestered carbon. We have no intention of focusing on a single trait in our selection program (e.g. K%), though the model in Chapter 13 does respond to K% in the sensitivity analysis.

Let me say a little about cultivars. A 'cultivar' is short form for cultivated variety, and is standard speak among those who've bred something new by crossing plants, or who've identified something in the wild they'd like to put a name to because of some useful characteristic – in the case of black walnuts it is so far the latter. In such a case, the name gives that particular genotype an identity, e.g. Emma K. Emma K is generally propagated by grafting some mature wood onto a non-select rootstock; a third or so of a mature tree will therefore not be Emma K.

Now, let me be unkind here and say that I believe the naming of black walnut cultivars has only been of benefit to the horticultural nursery industry. It has resulted in the sale of young trees at about $15-30 (no guarantees that you get what you pay for, for this can only be substantiated genetically), whereas a forest nursery will sell common (i.e. non-cultivar or unnamed) seedlings for $1-2 apiece if you buy them in bulk. Some workers have roughly calculated that planting cultivars can bring a nine-fold advantage in nut yield over nut production from plantations of the best common trees (if you knew which they were). My response to this is that the planting stock will still be cheaper if you plant nine times as many common trees. There may still, of course, be a yield disadvantage because the potential productivity of the common trees won't be known for some years, but then again I could say the same of any cultivar, if it is planted at a site far removed from its point of origin, where growing conditions may be significantly different.

If I am to accept that the cultivar/orchard model will work, I have to be assured that I will avoid costs associated with other models, while benefiting from increased returns. I believe I am correct in saying that that no-one has yet demonstrated this at the farm scale, largely because appropriately scaled technology doesn't exist. In fact there is a lot of work to be done in defining what this scale is. A pity those cultivar evaluations don't address NY.

In fact, I think some work should be done looking at NY of named selections on different rootstocks to gauge the 'rootstock' effect. After all, the hypothesis is that, if different rootstocks weren't necessary, the NY of any named selection should be uniform for a given DBH, all other factors (site, climate, on/off years, etc.) being equal. Not that I'm going to do it.

So, I don't work with 'cultivars', just >2,000 common trees, building the natural capital to let me exploit various income streams, and at least $45,000 avoided-costs in the bank.

As a footnote, in 2007 I made my first selections from the trees that nutted, and am now growing 25 F1 lines from seed. These will not be clones of the mother trees, as grafted cuttings would be, but until someone has done the science that convinces me that a $20 expenditure will make up for the reduction in heritability of the productive characteristics important to me as a grower in an industry which does not yet put value on any key trait of named cultivars[5], I prefer to select for traits which I have observed as useful here at the farm, rather than risk capital on Emma K, when it could turn out to be Humpty Dumpty.

Actually, I have some F1 Emma K also......

---

[1] http://www.agriculture.purdue.edu/fnr/HTIRC/pdf/woeste/Blackwalnut1959-1988.pdf

[2]

http://www.hammonsproducts.com/page.asp?p_key=75AB582D3D654ED699962F76AE09176C&ie_key=E4914FC71A964927B3783F585651E64E

[3] E.g. http://www.agriculture.purdue.edu/fnr/HTIRC/pdf/woeste/Blackwalnut1959-1988.pdf

[4] Though this interest may be academic if we can't show that nut-filling is influenced by disease expression.

[5] As money in the bank.

8. Potential Kernel Yield per Hectare

A function of energy partitioning and exogenous effects

Potential kernel yield per hectare is that variable which should be most closely allied to making or breaking BNP. Potential kernel yield speaks to the kernel composition of NY, but on a unit area basis rather than a singular focus on K% of the fruit. It speaks to the thinning strategy I have discussed in Chapter 6 which is meant to incrementally increase output by tree, and thus per unit area. It doesn't particularly matter what the starting average K% is as long as we work both to increase production and extraction. To do this we need to understand what the kernel is in energy terms and we need to discuss possible disease effects.

Black Walnut Energetics

I would define black walnut as a perennial oilcrop even if there is little scope for oil extraction (kernel extraction is hard enough!). Oils comprise 59% of kernel components, making the energetics of seed formation in black walnut an important focus. Energetics is the scientific study of energy flows and storages under transformation[1]. Nevertheless, it is important to understand overall tree energetics before examining energy partitioning into seed, so this is where we will start.

In Chapter 10 I elucidate a basic framework for simulation of growth and development of black walnut. The underlying argument is that as tree canopy, the site of photosynthesis, is the driver of growth (i.e. energy captured and transformed), what we do to that canopy must fundamentally affect growth outcomes. If we maximize open-grown canopy development through minimal pruning we ought therefore to optimize growth and development outcomes according to the tree's morphogenetic map. Because we may assume that the tree's own genetics will be oriented towards successful reproductive outcomes, we can leave the issue of the stimulus of physiological maturity and the thus the onset of nutting as itself canopy dependent.

In that chapter, I discuss the growth of black walnut as a very slender cone, and note that the growth rule underlying this morphology is replicated throughout the above-ground architecture. We could do (we haven't done them yet) some simple measurements showing the percentages of within-tree biomass laid down annually according to the diameter class of different tree components (bole, leading stem, major branches, minor branches, etc). We could similarly link root mass to above-ground biomass by allometric equations[2]. From here we ought similarly to be able to describe the canopy either as a surface or a volume, and assign it output coefficients corresponding to energy fixed per cm^2 or per cc, respectively. These coefficients would be derived from their expression in the physical biomass, the sink. In other words, by trial and error (another way of saying, depending on how many trees we measure in order to derive our coefficients) we should be able to come to some conclusion as to how much canopy is required to generate so much annual growth at our site. It is yet to be done.

Nut energetics

The nut is really a spectacular piece of biological engineering. Remember Figure3-1 where I showed the main physical parameters I use to describe the nut. Firstly, it is almost spherical, ensuring that for its volume energy expenditures per cc are minimized. Assume for the moment that evolutionary pressures within the species range have selected for optimum balance between shell and kernel. For this conversation I need to gloss over the fact that the internal structures of the shell are such that it is far from spherical. It is as if nature has worked out how to create lost-wax moulds into which it intends to cast the embryo. When the nut is newly harvested, every nook and cranny of that mould is occupied by embryonic material. As the nut dries, the embryo shrinks, separating itself from the shell walls, allowing us to see the bat-winged shape of the half-embryo (quite unlike the cerebral shape of the half-embryo of the Persian walnut). What we can be sure of is that that bat-winged shape is important to genetic survival, otherwise it would not have evolved. This is most likely something to do with energetics.

However, what I want to do for a minute is digress into observations on spheres, and their contained volumes, assuming also that the black walnut shell is laid down as a sphere, and that embryonic outcomes are a consequence of this. Let's take a rapid look at three spheres (Table 8-1), each double the diameter of first and second examples, respectively, and each with a 50% division of volume between shell and kernel cavity (very similar to what we saw in Chapter 3 where overall mean NIV was 0.497 of overall mean NEV). What is obvious at first is that as diameter doubles, so does total volume.

Diameter (cm)	Total volume (cc)	50% volume (cc)	Total energy density shell (MJ)	Total energy density kernel (MJ)	Total energy density of nut (MJ)
2	6.3	3.1	0.79	0.80	1.58
4	12.6	6.3	1.57	1.59	3.16
8	25.1	12.6	3.15	3.18	6.33

Table 8-1. Changes in volume (cc), and shell and kernel energy densities, with changes in nut diameter (cm), expressed for spheres.

If we then calculate the energy expended by the tree in forming the shell and embryo[3], this also doubles for each component as diameter doubles, though the relative energy density for shell and kernel volume is not exactly the same. There is considerable evidence that shell and kernel are not laid down simultaneously. All nutgrowers are familiar with 'floaters', physically fully-formed nuts (after husk removal) that float in water, but when cracked are generally found to have an aborted kernel. This is the simple test for seed viability that growers use – if a kernel is present the nut sinks. The aborted kernel generally takes the form of a blackened and shriveled testa with no evidence of other tissue.

From Table 8-1 it is not difficult to see that if the tree needs to put more energy into kernel density (the 'on' state of production), genetic manipulation of total nut volume, and the percentage of shell volume, allow the tree 'minimizing options'. A mechanism that switched kernel formation off in some nuts in times of stress, would enhance the probability of positive reproductive outcomes by ensuring viability in others. Observation suggests that the potential number of nuts per cluster on the tree is genetically controlled (i.e. each tree has its own average cluster size, quite uniform throughout its canopy but potentially open to environmental influences); in our plantations it varies from one to seven. Nut size does not seem to be particularly correlated with cluster size, though by the time we get up to seven-nut clusters[4], the nuts are generally smaller than the average across the plantation[5]. In multiple-nut clusters one can contemplate another option: that multiple nuts of total energy n MJ are likely to result in different reproductive outcomes than single nuts of the same total energy (i.e. the one-nut 'cluster', e.g. the bottom line of Table 8-1 compared with four times the top line). By reproductive outcome I mean relative presence (absence is a possibility)[6] of the mother tree's genes in the black walnut population on the landscape a generation 'later'. So, nut size and cluster size are ways of manipulating energy investments in reproduction.

Table 8-1 suggests that the difference in relative energy expenditures between shell and kernel (on an approximate 50:50 volumetric division) is small and probably not significantly different. It may be suggested that it would not appear to be to the genes' advantage[7] to increase kernel volume at the expense of shell volume (i.e. move towards a thinner shell), unless a concomitant move toward smaller total nut volume compensated for the shift in energy partition. Were there a large difference between relative energy densities of shell and kernel, different scenarios could be imagined, but they would have to take into account the pressure of co-evolution with our furry friend, the squirrel, i.e. under our scenario of equality, protection of embryos by shell is as important as vigour of the embryo itself. Though in the last section in Chapter 3, on nut genetics, I made the point that the energy cost of nut production suggests that an evolutionary strategy might be to test the growing conditions of the current year by pushing one or more shell dimensions to the maximum.

Shell Index

Thus I think we need to be able to identify trees where there is an expression of minimized investment in shell energy vis-à-vis kernel energy. None of the shell parameters mentioned so far (SW, and so S%) allow us to do this, as they have not incorporated the essence of this argument, which has to do with volume, i.e. energy packed into space. K% is similarly limited. For a given K% we have no idea of nut size. I suggest we use the concept of a Shell Index (SI), which would be calculated as g cc^{-3}, from SW/NEV. Any thin-shelled nut will express a low SI. Figure 8-1 illustrates the calculated SIs for the trees cracked in 2007. Most SIs are above 0.60, though I know there is a '3' tree in our stands with an SI much lower than this – it couldn't stand up to our traditional hulling methods and reliable data wasn't collected. Of named selections I have observed, I believe Emma K would similarly have a low SI.

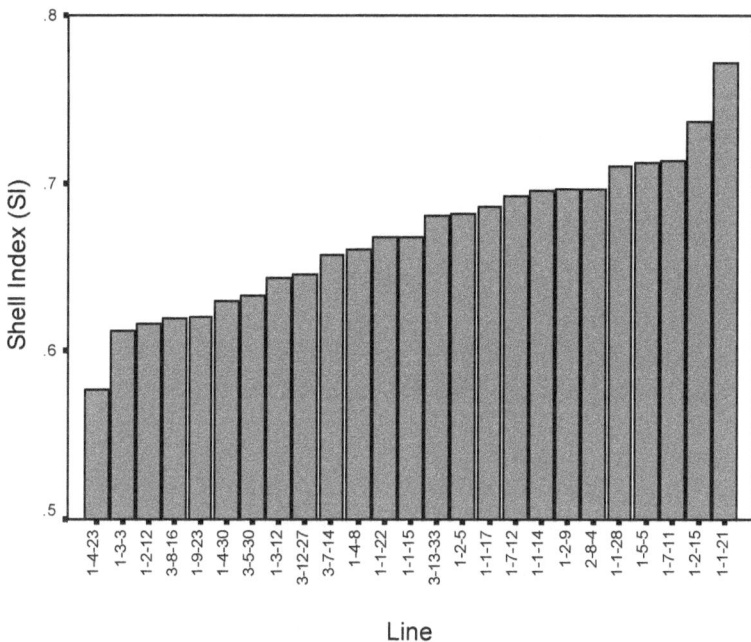

Figure 8-1. Individual SI in measured lines, 2007.

Disease

I have deliberately left disease issues until now. If there are substantial effects of disease on productivity, I think they will affect kernel production rather than NY (based on counts). I say 'if' because I have yet to see any. But let me go back and address the greater issue.

Disease control, and here I am talking principally about anthracnose and other leaf spotting disease effects, will be one of the most costly management operations that you can undertake[8]. I think it would eat up the profits of anything but the most productive enterprise. Therefore you need to be sure that you are improving your bottom line by doing it, rather than just improving the cosmetic appearance of your trees.

The first thing, of course, is to establish how susceptible your trees are to disease. The way we do this, again, is by scoring. By about early August, under our conditions, the relative differences in leaf-spotting disease expression, are quite evident, and run the gamut from trees which seem resistant to those which are hit hard. This is a colour expression, the former remaining dark green, the latter yellowing rapidly and losing leaflets throughout the canopy. I use the word resistant rather than tolerant because I think the degree of tolerance should be something we measure on potential kernel yield, i.e. if a hard-hit tree shows no decline in measured KY, we would ascribe this to tolerance. Resistance would be measured in changes of expression on the leaf.

Disease susceptibility cannot be understood without a systematic understanding of disease incidence throughout a plantation. I have scored all the trees in the two main fields discussed in this book for what I call disease range scores (DRS; Table 8-2). It is obvious that our disease scoring system, and thus the DRS, are quite different from the NRS. Firstly, our 0 in this case accounts for the few open spaces in our original planting pattern (i.e. trees absent). Then our DRS go from low to high in a reverse sense from the NRS - a high score is a negative feature rather than a positive one (this is quite by chance, though is perhaps the most intuitive way to work. It would be possible to have arranged the scores in the other direction). Thirdly, we do not attempt to separate diseases - we have lumped all disease expression in a single assessment, which is conducted when expression was evident.

Score	Count
1	No discernable damage (expression)
2	Few yellowed leaves
3	Moderate damage, tree lighter green, minor leaf loss
4	Significant damage, greater yellowing, medium leaf loss
5	Severe damage, very yellow, high leaf loss

Table 8-2. Disease Range Scoring System, Lostwithiel Farm.

The results for DRS are shown in Figure 8-1, by the distribution of trees across the DRS scoring range, for Field 3 in 2005 and 2007, and for Field

1 in 2006 and 2008. Scoring was conducted after August 1 in all cases, but earliest in 2005. Evident is that the majority of trees in both fields, for each year scored fall into the '4' category except for F3-05 where the majority is a '2'. All '1' and '5' score less than 10%, though all '1' really score less than 5%. This latter implies that there was a very small percentage which showed no leaf spotting or yellowing (high resistance), and equally that a small percentage showed high expression (low resistance). The majority '2' in F3-05 may be a reflection that disease expression increases with time (lateness) and that some trees might have been scored in a higher category had the scoring been done a couple of weeks later.

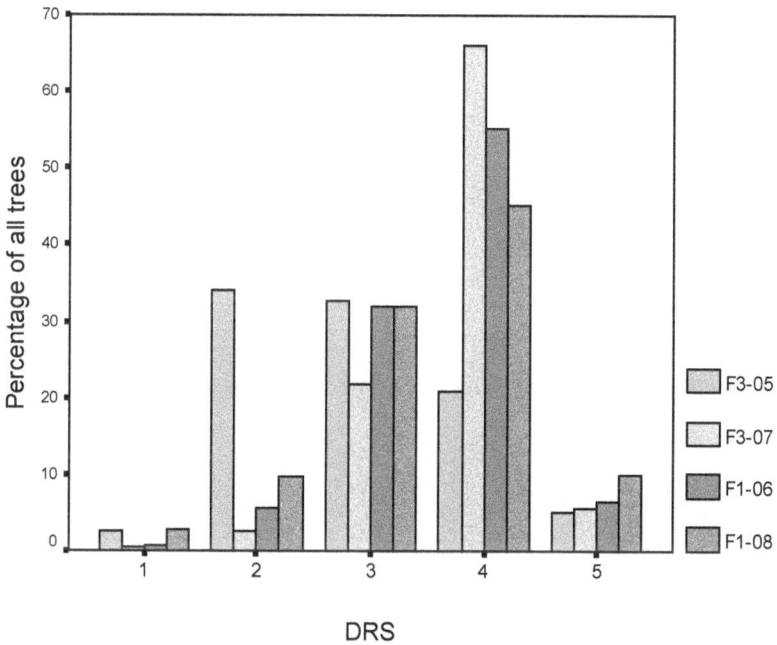

Figure 8-2. Percentage of trees in each DRS category, Fields 1 & 3, Lostwithiel Farm, 2005-2008.

These data can be examined in a several ways. First, we can look at the relative distribution across all range scores, as we have done above. Secondly, we can look at the pattern within a field across years, to see if the expression, on average is different between years, and thirdly we can look at the DRS scores of the top trees in our line index (Chapter 12) to see what we can conclude.

The extent to which trees score consistently (or otherwise) for DRS only matters if there is some correlation between DRS and NY. So far let me just say that the evidence for this is minimal, though two '5 DRS' trees do appear in the full line index for '3NRS' trees [A][9]. Mean DRS for the two fields only changes by 0.10 between the first and second set of observations, though this is principally due to the scores of Field 1 trees .Field 3 '3' trees returned identical DRS scores between 2005 and 2007 [A].

Conclusions

Where has this discussion taken us? It has highlighted the basis for energy accumulation in the nut, and that the key management strategies will emphasize maximum captured energy output per unit area. As nearly all our trees occupy the same site area, it is not hard to see that NY is as good an index as any other, and that ways to increase NY on a unit area basis are strategies that will serve us best. Thinning to maximize NY in the remaining trees is one of these. Disease scoring is an aid to management suggesting that we do not need to spray for disease control as there appears so far to be little reflection of DRS on NY, nor an accumulation of disease pressures across years.

----

[1] Wikipedia.

[2] Where someone else has already done the work and calculated the functional and/or structural ratios.

[3] I have used 19.25 MJ /kg for shell, and 23.01 MJ /kg for air dry kernel, both corrected for 15% moisture. The former was determined for me; the latter derives from the USDA National Nutrient Database Shell specific gravity is about 1.3g/cc; dry kernel is very similar, but at the moment it is laid down, green kernel specific gravity would be less – I have used a value of 1.1.

[4] In fact, I've only discovered a single tree with this cluster size, in 2007, and it did not show the same trait in 2008.

[5] This is an anecdotal observation; we still have to confirm this by measurement

[6] i.e. gene frequency.

[7] Here the point is that we are not talking about a 'tree' or 'species' advantage - the tree and species are but the particular vehicles by which the genes achieve their replication.

[8] A similar remark could be made about pest/insect control. As we currently only see sporadic attacks by the walnut caterpillar, we feel that it would be more costly to spray for control than bear the consequences of no control.

[9] Appendix A.

9. Net Kernel Yield per Hectare

A function of potential kernel percentage and extractablility

This chapter I have deliberately left to a future edition of this book, if feasible, because I believe that net kernel yield per unit area will be driven as much by how, mechanically, we can crack and separate than by anything else. If we manage, through stand management, to maximize potential kernel yield per unit area, it stands to reason that processing efficiency ultimately defines our monetary return. The corollary is that our potential kernel yield per unit area will tell us how much we can afford to invest to achieve the efficiency necessary to make a profit on net kernel yield (some of this is dealt with in Chapter 13).

I can hear a collective groan of dismay that this book doesn't cover processing mechanization. The simple reason for this is that I could not yet do it justice. I have had a collaborative program of technology development ongoing with a college partner for the past five years and Figure 9-1 summarizes where I think we are in that process. But to analyze the mechanics of processing to the same level of detail as BNP will take an accumulation of data over the next five years. Let me say nothing more than I hope that I can do it.

Figure 9-1 shows our stage of development for each of the component parts of our processing scheme. We have deliberately chosen to start from scratch for reasons of cost, farm-scale and potential productivity of black walnut under BNP. There are three levels leading to a commercially ready machine, with Figure 9-1 showing for each machine where we are in its development. Proof-of-concept is the version that comes off the drawing board. The functional prototype is the workable proof-of-concept. A commercially-ready machine should have had all of the prototype's kinks worked out.

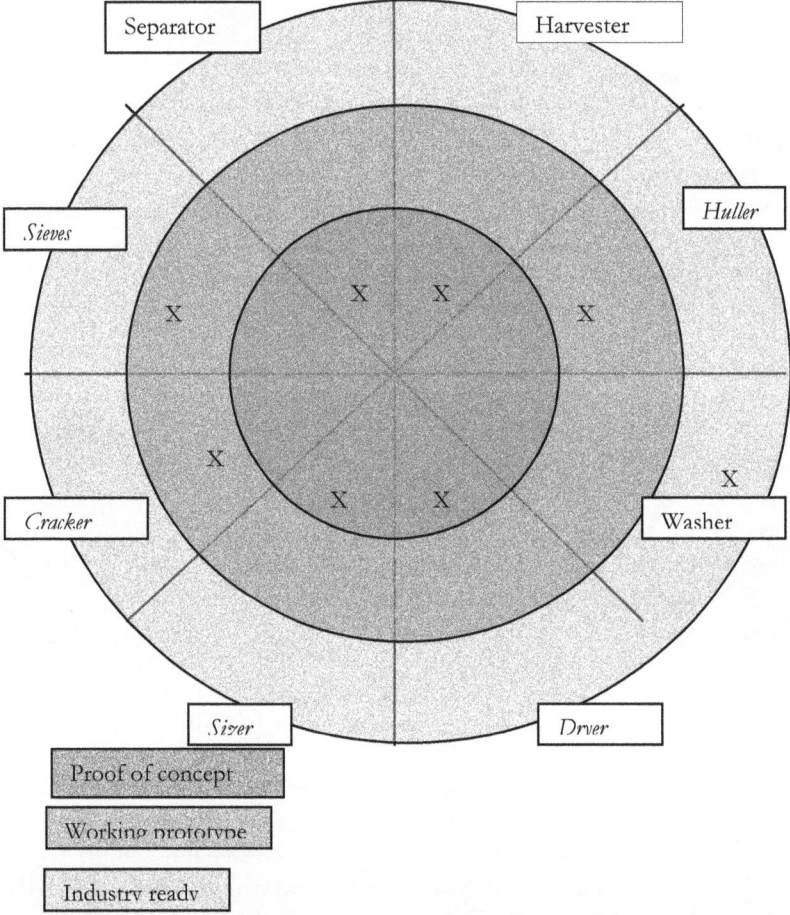

Figure 9-1. Progress in collaborative development of the processing technology suite.

This program has run in parallel with the development of physiological maturity in our oldest BNP stands. Obviously not a perfect match, but we have done our best. Are we heading towards marketing commercially-ready machines? Only time will tell.

10. Carbon sequestration with black walnut

A function of tree size and proportional distribution of carbon

One of the basic questions I started asking myself several years ago was: In this era of fossil-fuel dependency, can BNP contribute significantly to carbon sequestration? In fact, the question arose when the conical growth rule first surfaced; I began to see how I could easily quantify C sequestration (storage). Many species have been found to exhibit constant (not equal) relationships in the growth of different parts, and this allometric growth, as it is called, helps us to estimate growth in that part of the tree we can't see, the roots. In actual fact, we'll go some way to calculating a tree equivalent for carbon offsetting strategies, as much is talked about this, some of it dangerously close to snake oil equivalents. Our carbon debt is large and growing.

But we are going to begin this discussion by looking at the change in the two variables that we will use to compute volume: H and DBH. The pattern of annual change in these variables in Field 3 is shown in Figure 10-1. I have expressed DBH as DBH*100 in order to plot it onto the same vertical axis as H. Here we see that both H and DBH are increasing, H by about 100 times as much as DBH! It is important to remember that DBH itself is not increasing at this rate so the relative slope of the DBH line (compared to that of DBH*100) would be much less.

Conical (Growth) Rule

It is possible to think of our almost 600 annual DBH measurements on trees of varying diameter class in Field 3 in another way: as measurements taken of diameter of an average tree at varying distances from the growing tip, i.e. with reference to the top of the tree, not the bottom. This data is shown in Figure 10-2. Our new analysis generates a straight line. Not only this, but that effectively the same straight line is valid for different years (see below). What does this mean?

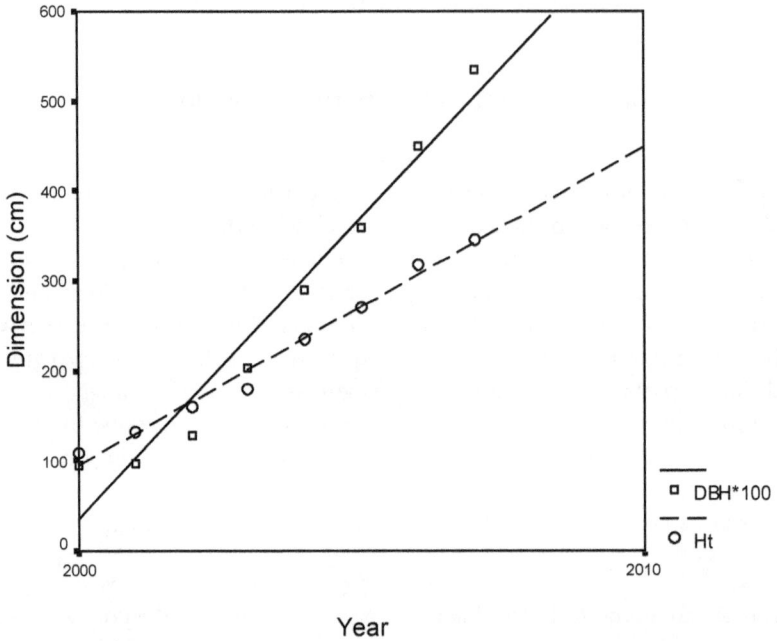

Figure 10-1. Pattern in H and DBH*100, Field 3, Lostwithiel Farm, 2000-2007.

It means that our average tree is growing as an extremely elongated cone, and that the same depth of wood is laid down in absolute terms at 10 cm from the growing tip as at 3 m from the growing tip, or at the point at which DBH was 'measured' on this average tree. All of a sudden we see that DBH is a different beast than pure D (lets call this D at DRT - distance relative to tip). The fact that annual DBH measurements were effectively made at different points on the lines expressed in Figure 10-2 confounded (mixed up) true diameter change with this other measurement that related more to human height than to tree height.

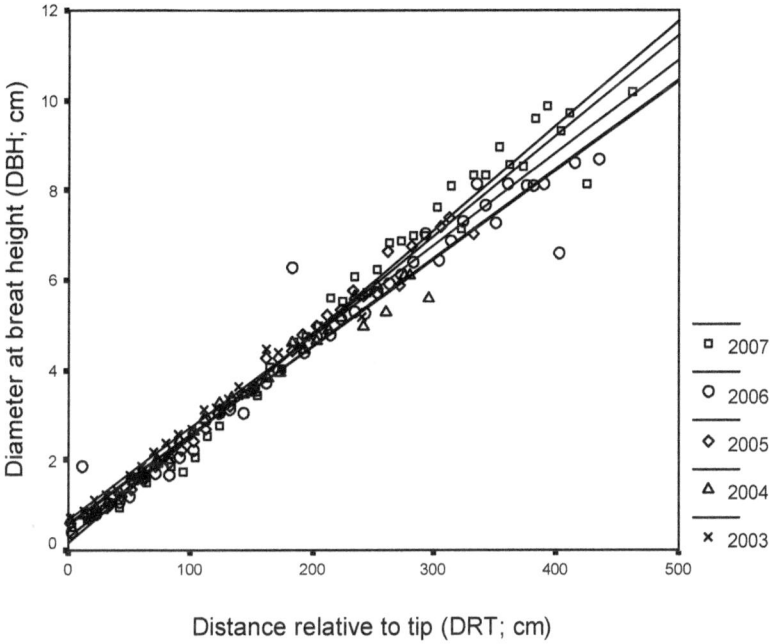

Figure 10-2. Relationship between DBH and DFT, Field 3, Lostwithiel Farm, 2003-2007.

Let's clarify one thing, however. Even though we have new cones of exactly the same characteristics (from here on, our conical rule) being laid down each year, relative diameter change at different DRTs is still different, i.e. 0.5cm laid down at a DRT of 1 m yields a different RDGR (0.19) than 0.5cm laid down at a DRT of 3 m (0.07). Now, however, we know that our RDGR for a given DRT is a much truer expression of annual growing conditions because we have removed the confounding of DBH from our assessment. Figure 10-3 gives you a visual sense of this conical expression

In other words, black walnut grows as a very slender cone but not with increasingly greater relative D at the base as interpretation of DBH would imply. Wood is laid down annually at almost precisely the same depth at all 'heights' on the tree (there is some minor variability, but it is this consistency which makes a black walnut look like a black walnut).

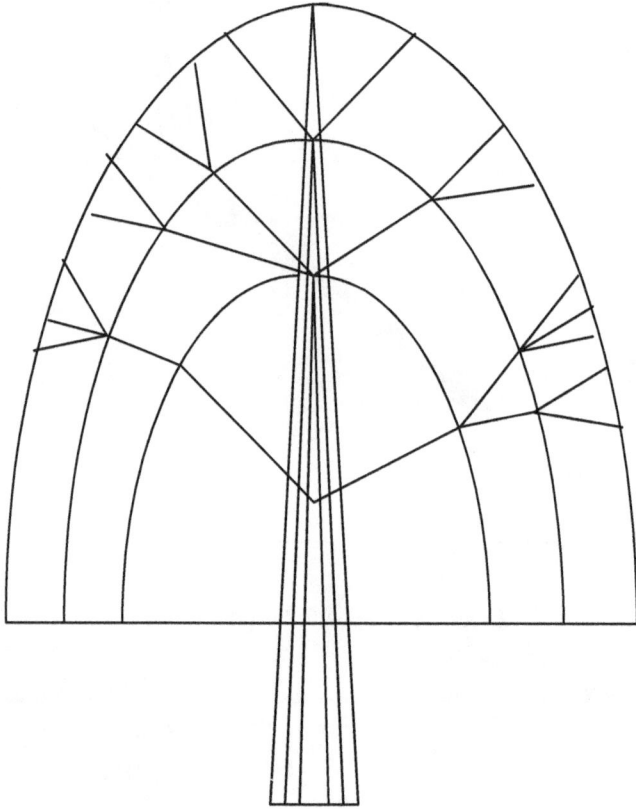

**Figure 10-3. Two-dimensional representation of open-grown
black walnut incorporating key growth features
(lowest stem cones and branches not shown)**

The depth added by year is defined by the amount of vertical increase in
H in that year (alternatively, the amount of vertical increase in height in
any year is defined by the conditions which allow the tree to generate
wood). If we try and visualize how the tree is developing, we can start
with the idea of an over-sized helmet (the canopy) on a hatstand (the main
stem). Each year the hatstand increases in height and girth, and another
helmet is laid down over the first. It may or may not be of similar thickness to
the previous one, but its shape will be derived from the first helmet's shape.
The reason for this is that all branches and twigs within a tree will obey the
conical rule as expressed by that tree. Thus, if the helmet is of hemispherical
shape from an early year, the form of the canopy will not change markedly in
the coming few years. It will just expand.

At the surface joining successive helmets, there is ongoing furcation of branches, all according to the probabilities that govern the number of buds which elongate. As we've noted, branches grow according to the conical rule (not shown in Figure 10-3), for there is no morphological difference between our main stem and its various branches other perhaps than apparent size (i.e. they all look as though they belong to a black walnut. If we were to take measurements, we would find the same relation expressed from the tip and down to the start of each branch's ring collar. I haven't yet done it, but I am confident enough of my eye to guarantee it).

Regression analysis indicates that annual variability is not a significant factor in our analysis of Figure 2[1]. If we pool (combine) the data, we obtain a regression equation of: D = 0.511 + 0.02085DRT, i.e. at each successive cm from the tip the D increase is just over 0.2mm. The actual measurement is calculated by adding 0.511 to 0.02085 x chosen DRT, e.g. at 1m DRT, D measures 2.596.

This relationship allows us to do a couple of things:

- calculate the angle expressed at the tip of the cone; and
- estimate an annual height increment if we were to know RDGR at uniform DRT. This is not quite so easy, because we are actually using DBH and H to calculate the shape of the cone. However, we can turn it around and say that if we were to encounter an Xcm annual increment in D at a previous year's DRT of 1m, our H increment would be Ycm.

Calculation of the conical angle is simple: from school-level trigonometry we remember SOHCAHTOA for right-angled triangles. Our formula above tells us that at a DRT of 1m, D will be 2.596cm. The DRT is our measurement of O, and D is 2xA, so we use the TOA part of SOHCAHTOA and calculate the value of the tangent of half of our conical angle:

$$\frac{2.596}{2 \times 100}$$

= 0.01298 radians or 0.744°. The angle at the apex of our cone is twice this, or 1.487°.

Then, if we marked this DRT, and found it had, for example, increased to a D of 4cm, we would find our H increment estimated to be 54.1cm.

This brings us full circle in our valuation of DBH as a useful index. In a sense, DBH is permanently marked, so we could always estimate annual increments in H from annual increments in DBH, but only once we had already derived our conical rule. Statisticians are always nervous about using an equation such as that above to extrapolate either outside the period (another year) or range (in DBH or DRT) measured, for they have no proof that things will always stay the same. Given that we have 5 years of consistent data, and my trees still look like black walnuts (which they wouldn't necessarily do were the relationship to change), I am fairly confident that it can be used for extending calculations beyond this period (not that I am going to do this here).

The conical growth rule was calculated from the observations we took of DBH and H on all our trees in Fields 1 and 3 until 2007 (Field 1 from 2006: the conical rule was consistent with that in Field 3, further adding to my level of confidence in it.

A critical assumption is the division between above-ground and below-ground biomass. A thorough search of the literature to date does not provide a reliable estimate of black walnut shoot:root ratios, especially in growing plantations, so I have chosen to assume that below-ground biomass is one third of that above-ground (a rough approximation of what the literature suggests). This may eventually be disproven.

Table 10-1 combines all these relationships to demonstrate the amount of C we are sequestering annually. A feature to note here is that I am using C (i.e. actual carbon), not CO_2 equivalents, as my means of measurement. I think there is something ethereal about CO_2 which confuses how much actual C is being sequestered.

	2003	2004	2005	2006	2007
Total volume of central stems (cu m)	1.5	2.4	3.3	5.0	7.3
Total no of trees measured	466	507	525	534	542
Number of trees per ha	277.8	277.8	277.8	277.8	277.8
Volume of stems per ha (cu m)	0.9	1.3	1.7	2.6	3.7
t C per ha (tops, i.e. stems)	0.5	0.7	1.0	1.5	2.1
t C per ha (belowground)	0.2	0.2	0.3	0.5	0.7
Root:shoot ratio	0.33	0.33	0.33	0.33	0.33
Total t C per ha	0.6	1.0	1.3	1.9	2.8
Annual Increment (t C per ha)		0.32	0.33	0.64	0.86
Annual % increase		49.6%	34.2%	49.0%	44.5%

Table 10-1. C sequestration rates, Field 3,
Lostwithiel Farm 2003-2007.

Quite obvious from Table 10-1 is that we do not appear to be sequestering significant amounts of C, though this is rapidly increasing with time. We can confidently say that it is now in the range of tonnes per ha, and not anything less. It should be noted that calculations are made on the basis of the volume of the central stem, and thus do not include branches, so are conservative estimates of the amount of carbon sequestered. The field was planted in the 1990s, so the figures in the first column represent cumulative values prior to the year indicated.

Until landowners are paid for ecosystem services rendered, this exercise is largely an academic one. However, this example demonstrates how the calculations can be done, and the magnitude of the amount of C sequestered. At a carbon price of $30/t, black walnut at this age and under our conditions would yield annually approximately an extra $10/ha for sequestration alone.

Let's take it a step farther and examine a trial farm carbon budget. In Table 10-2 I have approximated on-farm energy usage, and so our carbon footprint. At 2007 values these two fields do not offset our annual carbon debt (they only reach about 25% of it). It is probable that were we to include other on-farm tree blocks we would achieve full offset, but again I have wanted to keep the calculation clear, and based on the trees that we currently measure. So we can conclude by saying that in the early years we are hard put to make significant inroads into carbon debts by offsetting schemes, and it would only be once we started to think of life-timeframes that we would get into positive outcomes. Something for offset planners to think about.

Year	2007			
Debit				
Item	unit	amount	C/unit (kg)	total C (t)
Vehicle fuel (gas)	litres	2500	0.62	1.55
Tractor fuel (diesel)	litres	200	0.73	0.15
House, propane	litres	1000	0.62	0.62
All buildings, electricity	kwh	17000	0.12	1.97
Total				**4.29**
Credit				
Item	unit	amount	C/unit (t)	total C (t)
Black walnut, field 1	ha	3.12	0.14	0.44
Black walnut, field 3	ha	1.95	0.32	0.63
Total				**1.06**

Table 10-2. Farm Carbon Budget, Lostwithiel Farm, 2007.

The final question: How many trees to a tank of gas (petrol)? As I said in my blog back in November, 2007: What I really mean by this is, how many trees of the average age and size of the trees we have in Field 3 will it take to sequester the carbon emitted by combustion of 50 litres of gas? The answer is, as of 2007, about 10. You can see (Table 10-3) that the trees are growing quite rapidly, increasing in sequestering capacity by about 50% per year (except 2006, which was very wet and cool, and apparently not favourable for growth; a substantially thinner cone was laid down in 2006 than 2005). Annual estimates are made on the increment in sequestration capacity, and not on the basis of total carbon fixed.

	2003	2004	2005	2006	2007
Average C per tree (kg)	2.33	3.48	4.67	6.96	10.05
Annual C increment per tree (kg)		1.15	1.19	2.29	3.10
Annual C increment per tree (%)		49.6%	34.2%	49.0%	44.5%
Litres of gasoline equivalent in increment (l)		1.86	1.92	3.69	4.99
Tree equivalents per 50 litre tank of gas		26.87	26.04	13.56	10.01

Table 10-3. Tree equivalents in a 50 litre tank of gas (from yearly increment), Field 3, Lostwithiel Farm 2003-2007.

[1] The purists among you may say: I can see daylight between the lines in Figure 10-2, so they cannot be identical. In response, I am not saying they are identical, just that the variability between years is so minor, that the daylight can be ignored. Much of it may be due to distortion by the outliers at higher values of DRT, of which there are few.

11. Black walnut as a functional food[1]

A function of whatever else we eat

Nuts generally are receiving increasing attention in terms of the possible health benefits regular consumption of even small amounts may bring. In this context they can be described as functional foods (e.g. Hasler, 2002), i.e. foods that deliver more than just the expected nutrition which a proximate analysis suggests they should. Recent evidence of this attention was the petition from the California Walnut Commission (CWC) to the United States Food & Drug Administration (US FDA) to allow a claim of a health benefit on packages containing the Persian Walnut, *Juglans regia* (FDA, 2004). The fact that the FDA disallowed the principal request did not negate the claim; rather it spoke to the lack of studies which prove conclusively that it is walnut, and walnut alone, which delivered the claimed benefits. Nevertheless, the evidence is mounting for an eventual understanding of the true value of nut consumption to human health.

The FDA now notes (2008)[2]: *Scientific evidence suggests but does not prove that eating 1.5 ounces per day of most nuts as part of a diet low in saturated fat and cholesterol may reduce the risk of heart disease.* Qualified health claims tend to be associated with specific constituents rather than whole foods, though Persian walnuts appear to be an exception. A summary of potential health benefits is included in Table 11-1. A common framework is emerging between many countries about what may be claimed and how[3].

Black walnut growers have a vested interest in this issue. As much of North America is inhospitable to *Juglans regia*, if it were shown that *J. nigra* brought comparable benefits, marketing strategies for the latter would gain an overnight boost. However, basic questions relevant to comparisons of the two species are: Which constituents bring the health benefit, and is the relative content of these constituents between species important to the analysis?

Constituent	Potential Health Benefit
Unsaturated fatty acids	Help reduce total blood cholesterol, improve cardiovascular condition, improved longevity
Vitamin E	Antioxidant, reduced cardiovascular disease risk, reduced cancer risk
Arginine (including low lysine:arginine ratio)	Improved platelet function, reduction in atherosclerosis, precursor of nitric oxide synthesis which influences vascular tone
Phytosterol	Cholesterol reduction, anti-microbial activity
Folate	Reduced risk of cardiovascular disease (independent of cholesterol-related issues)
Minerals, e.g. Mg, Cu, Se, Zn	Blood pressure, enzyme function, tooth enamel
Dietary fibre	Cholesterol reduction

Summarized from Feldman (2002) and Kris-Etherton et al. (1999).

Table 11-1. Potential health benefits of walnut consumption, by main functional constituent

Proximate Analyses

An analysis of the main nutritive characteristics of three *Juglans* species is shown in Table 11-2. Butternut (*J. cinerea*) and pecan (*Carya illinoensis*) are included for comparison. What is notable, in the first instance, is that the proportion of fat, per 100g edible portion, is quite variable between species (from 57% in butternut to 72% in pecan). However, the protein content varies even more, from 9% in pecan to 25% in butternut. Black walnut is very similar to butternut; Persian walnut is intermediate. Carbohydrates are the least significant component of the three main nutritional groups; starches are an insignificant percentage of the carbohydrates.

The first important observation is that this data is still quite limited, and the United States Department of Agriculture (USDA) National Nutrient Database from which it is drawn is not consistent between species in the number of samples analyzed, especially for black walnut and butternut.

	J. nigra	J. regia	J. cinerea	C. illinoensis
Water (g)	4.6	4.1	3.3	3.5
Energy (kcal)	618	654	612	691
Fat (g)	59.0	65.2	57.0	72.0
Protein (g)	24.1	15.2	24.9	9.2
Fibre (g)	6.8	6.7	4.7	9.6
Carbohydrate (g)	9.9	13.7	12.1	13.9

Source: USDA National Nutrient Database (2009); http://www.nal.usda.gov

Table 11-2. Proximate analyses of J. nigra, J. regia, J. cinerea and C. illioensis

It is known, for Persian walnut, that fat content alone can be quite variable between selections (Savage, 2001, range 62.6-70.3%; Amaral et al., 2003, 62.3-66.5%). In Savage's study, protein content ranged between 13.6-18.1%, and dietary fibre from 4.2-5.2%. Wakeling et al. (2001), however, found no differences in a wide range of constituents in two pecan cultivars. USDA sample sizes for proximate analyses of black walnut and butternut are small, and the more detailed analyses of fatty acid and amino acid profiles, and other potentially bioactive constituents, are often single observations (i.e. $n=1$). In the first instance, then, future research in this area in all species requires significant expansion in range of material analyzed in order to account for the percentage range in constituents. It is possible that intraspecific ranges may stray into the extremes of interspecific contrasts. It is not immediately known whether the study of Wakeling et al. (2001) dealt with two cultivars of close genetic background.

Other Nutritional Characteristics

Some values for selected nutritional characteristics are shown in Tables 11-3 and 11-4. Their selection is based on a consensus in the nutrition literature, reflected in Table 11-1, that they are probably the key constituents, in terms of known bioactive compounds. The arguments for this are extensive, and it is not proposed to repeat them here - readers are referred to Sabaté (1999), Kris-Etherton et al. (1999) and Feldman (2002) for further reading on this topic. These papers can be downloaded from the internet via PubMed. The purpose of the present paper is to highlight where, based on an initial analysis of the USDA data, black walnut might stand in relation to the Persian walnut in delivering the principal constituents of a health benefit, and, hopefully, the benefit itself. Table 11-3 is expressed as content per 100g edible portion; Table 11-4 is expressed as ratio of content per 100g edible portion between *J. nigra* and each of the other three species.

Common understanding of diets and nutrition says that fats are undesirable elements of daily food intake as they lead to weight increases, cholesterol deposition, and diseases (especially cardiac illnesses) which result from fatty diets. In fact, part of the rationale for the FDA rejection of the CWC petition was based on the high total fat content of walnuts. However, the type of fat consumed plays a significant role in such diseases, and it is not enough to talk just of fat alone. Close analysis of fat contents in nuts shows quite significant variability between the main three fat groupings (saturated, mono- and polyunsaturated). In general, the lower the level of saturated fats (SFA) the better, especially as a ratio to monounsaturated (MUFA) and polyunsaturated (PUFA) fats - the A in these abbreviations refers to Acid, as these fats are normally referred to as *fatty acids*. The four examples in Table 11-3 show a high proportion of MUFA and PUFA in relation to SFA, especially in butternut, the range being somewhere between 10-20 times. It is partly this ratio which appears to be of importance in terms of the health benefit of nuts.

	J. nigra	J. regia	J. cinerea	C. illinoensis
Saturated fat (SFA, % of fat)	5.7	9.4	2.3	8.6
Monounsaturated fat (MUFA, % of fat)	25.4	13.7	18.3	56.7
Polyunsaturated fat (PUFA, % of fat)	59.5	72.3	75.0	30
Linoleic (omega-6, % of PUFA)	94.3	80.8	78.9	95.4
Linolenic (omega-3, % of PUFA)	5.7	19.2	20.4	4.6

Source: USDA National Nutrient Database (2009)

Table 11-3. Relative concentrations of main fatty acid constituents in J. nigra, J. regia, J. cinerea and C. illinoensis

To show the nature of comparisons of interest, the example of alleged total cholesterol lowering effects will be considered. In Kris-Etherton et al. (1999), the authors examined many human dietary studies where fat consumption had been measured, and where responding blood cholesterol levels had also been measured. Predictive equations typically of the following form were generated from the data to relate changes in blood cholesterol to fat consumed:

$$\text{Change in total cholesterol} = a\,(\text{change in SFA}) - b\,(\text{change in MUFA}) - c\,(\text{change in PUFA})$$

This equation states that from the beginning condition, where the test person is consuming a known diet, if the proportional changes in consumption of SFAs, MUFAs and PUFAs are known, the changes in blood cholesterol can be predicted. Some previous studies (Mensink and Katan, 1992, cited by Kris Etherton et al., 1999) had calibrated this equation to provide values of the constants a, b and c where dietary change was from a diet where energy was all from carbohydrates and free of fatty acids, to one where some energy was replaced by SFAs, MUFAs and PUFAs in varying proportions. What is important in this equation is that both MUFAs and PUFAs have a negative sign, which means that they lower blood cholesterol, but proportionally in relation to the consumption of SFAs. (SFAs themselves increase blood cholesterol levels by more than the consumption of dietary cholesterol.) The relative proportions of the fatty acids of each nut species are shown in Table 11-3. To calculate a relative beneficial effect from nut consumption by species I have assumed that some of the carbohydrate in the diet has been replaced by 100g of nuts. As these are relative values, we can express calculated potential species effects on total cholesterol reduction relative to each other.

If all values are adjusted to compare to a baseline value health benefit of 1.0 for black walnut, then butternut, Persian walnut, and pecan give proportional benefits of 1.1, 1.33 and 0.74, respectively, i.e. butternut, Persian walnut and pecan should be 10% more, 33% more and 26% less effective than black walnut in lowering total cholesterol, respectively. However, this interpretation is highly speculative, and it is possible that these differences are not large enough to be statistically significantly different.

The omega-3 (alpha-linolenic acid) and omega-6 (linoleic acid) fatty acid composition is also thought to be linked to the health benefit (Kris-Etherton, et al, 1999). Both are essential elements of the diet, as the body cannot synthesize them. Table 11-3 shows the omega-3 and omega-6 balance in the PUFAs of each species. Within the limits of the data, there appear to be two groups: that where the omega-6:omega-3 ratio is 19:1 (black walnut, pecan), and that where it is 4:1 (English walnut, butternut). If these data are representative of each species, then to meet a daily recommended intake of omega-3 fatty acids, taking into account the proportional differences of PUFAs in each species, would require for men and women, respectively, the consumption of 18 g and 13 g of English walnut or butternut, and about four times as much black walnut or pecan. However, as the literature suggests that pecans also bring some of the benefits known for Persian walnut (Morgan and Clayshulte, 2000), either the sample data shown here is misleading, or other constituents of nuts are as important as the omega-3 fatty acids. Other bioactive constituents of nuts include protein, dietary fibre, and micronutrients such as copper and magnesium. Nuts are also a rich source of Vitamin E and arginine. It should be noted that the quantitative analysis of cholesterol-reduction mentioned above, when applied to known studies of nut consumption, indicates that the actual benefit exceeds the predicted benefit by some 25% (Kris-Etherton et al., 1999), i.e. it is highly likely that other compounds are also contributors. One must also ask whether effects are just additive, or more likely, that there are interactive effects between constituents.

It is not yet possible to undertake for non-fat constituents the type of analysis done with unsaturated fat effects on cholesterol. Instead, Table 11-4 shows the proportions of each constituent per 100g edible portion in the different species adjusted to the their value relative to black walnut. Across the range of constituents, black walnut compares well with the other species in constituent concentrations; the notable exception are folate and zinc. Again it is important to note the current limit of this analysis imposed by the number of sample points.

	J. regia	*J. cinerea*	*C. illinoensis*
Arginine	.63	1.35	.33
Lysine:Arginine ratio	.95	.80	1.20
Beta-sistosterol	.62	n.a.	.86
Folate	3.16	2.13	.71
Magnesium	.79	1.18	.60
Potassium	.86	.82	.80
Zinc	.91	.91	1.32
Copper	1.14	.36	.86
Manganese	.87	1.69	1.15
Selenium	.29	1.01	.22
Vitamin E (α tocopherol)	.39	n.a.	.78
(γ tocopherol)	.73	n.a.	.86

Calculated from values in USDA National Nutrient Database (2009); n.a. = not available; *J. nigra* = 1.0 for all parameters.

Table 11-4. Values of selected nutritional parameters in J. regia, J. cinerea and C. illioensis expressed as a ratio of their presence in J. nigra (100g samples).

The second important observation, both on the data and on the analysis, is the importance of shell percentage in terms of what might be termed the marketable yield of health-imparting and/or bioactive constituents. All the data presented so far have been on the basis of concentrations per 100g sample of kernel (except for percentage values in fatty-acid profiles). If the three main constituents of the proximate analyses, carbohydrate, protein and fat, are represented in terms of approximate proportional content in nuts of equal weight (Figure 11-1), it may be seen that species of higher shell thickness have proportionally lower fat contents, though protein content is reasonably uniform [approximate shell percentages were derived from: pecan, Madden (1979); Persian walnut, CWC (2004); black walnut, Thomas (unpublished data). Other constituent proportions correspond to the mean values from the USDA National Nutrient Database (2009)]. The simple and perhaps not inappropriate assumption is that shell and fat are the two high-energy sinks in the nut, and that interspecific differences in energy deposition reflect niche-driven selection outcomes; an equivalent simple assumption is that protein, as the biological machinery of the nut, provides essentially similar services across species, and that its approximate uniformity (in overall not relative terms) should not be unexpected. However, the higher shell percentage of black walnut, compared to Persian walnut and pecan, means that a greater weight of total nut must be processed to extract an equivalent volume of kernel. If, in the end, there is an advantage in marketing black walnut on the basis of its health-delivering constituents, much more will need to be known about relative content of these compounds in currently-grown selections and in the natural population at large.

If the Persian walnut provides a range of health-giving and bioactive constituents, black walnut almost certainly does the same. Many references are made of the importance of the black walnut, among other species, to the past diet of American aboriginal people[4].

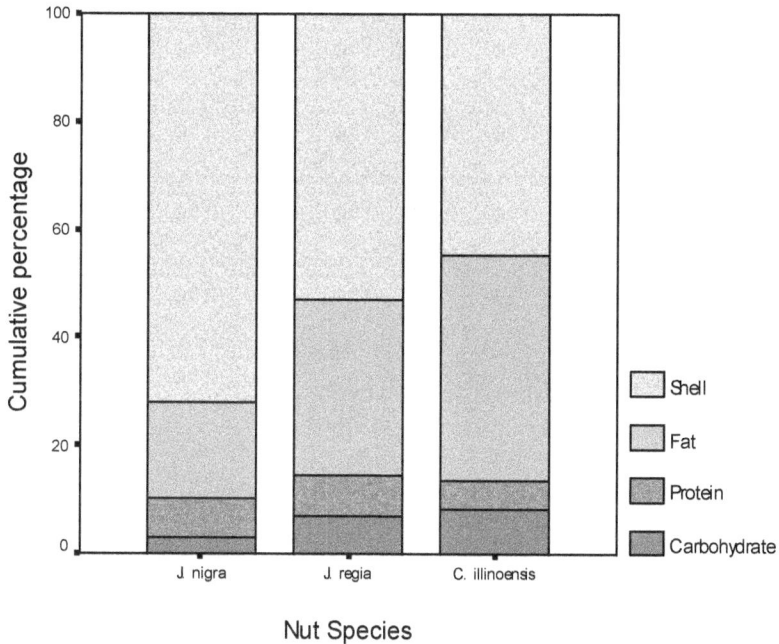

Figure 11-1. Approximate content of main whole-nut constituents, by nut species

The US FDA granting of a qualified health claim for the Persian walnut is an important first step in recognition of the functional food value of nuts, and will stimulate more research in this field, though pitfalls still lurk[5]. But black walnut producers will not automatically be able to ride on the coattails of permitted health-benefit claims for the Persian walnut. Science will be making the determination of the effect at the constituent level, and species-specific knowledge will be essential to any future granting by the US FDA of equivalency between nut species in delivery of health benefits.

Literature Cited

Amaral, J.S., S. Casal, J.A. Pereira, R.M. Seabra and B.P. Oliveira, 2003. Determination of sterol and fatty acid compositions, oxidative stability, and nutritional value of six walnut (Juglans regia L.) cultivars grown in Portugal. J. Agric Food Chem, 51(26):7698-702

California Walnut Commission, 2004. http://www.walnuts.org

FDA, 2004. Qualified Health Claims: Letter of Enforcement Discretion - Walnuts and Coronary Heart Disease. Docket No 02P–292,Office of Nutritional Products, Labeling and Dietary Supplements, Center for Food Safety and Applied Nutrition, Dept of Health and Human Services, U.S. Food and Drug Administration, March 9, 2004

Feldman, E.B., 2002. The Scientific Evidence for a Beneficial Health Relationship between Walnuts and Coronary Heart Disease. LSRO Report. J. Nutr. 132: 1062S-1101S.

Hasler, C.M., 2002. Functional Foods: Benefits, Concerns and Challenges - A Position Paper from the American Council on Science and Health. J. Nutr. 132:3772-378.

Kris-Etherton, P.M., S. Yu-Poth, J. Sabaté, H.E. Ratcliffe, G. Zhao and T.D. Etherton, 1999. Nuts and Their Bioactive Constituents: Effects on Serum Lipids and Other Factors that Affect Disease Risk. Am J Clin Nutr. 70 (Suppl):504S-511S.

Madden., G. 1979. Pecans. *In*: Nut Tree Culture in North America. *Ed*: Jaynes, R.A. NNGA Publication, 13-34.

Mensink, R.P., and M.B. Katan, 1992. Effect of dietary fatty acids on serum lipids and lipoproteins. A meta-analysis of 27 trials. Arterioscler Thromb. 12:911-9.

Morgan, W.A., B.J. Clayshulte, 2000. Pecans lower low-density lipoprotein cholesterol in people with normal lipid levels. J Am Diet Assoc. 100(3): 312-8.

Sabaté, J., 1999. Nut Consumption, Vegetarian Diets, Ischemic Heart Disease Risk, and All-Cause Mortality: Evidence from Epidemiologic Studies. Am J Clin Nutr. 70 (Suppl):500S-3S.

Savage, G.P., 2001. Chemical Composition of Walnuts (*Juglans regia L.*) Grown in New Zealand. Plant Foods Hum Nutr. 56(1):75-82.

USDA, 2004. National Nutrient Database; http://www.nal.usda.gov

Wakeling, L.T., R.L. Mason, B.R D'Arcy, and N.A. Caffin, 2001. Composition of pecan cultivars Wichita and Western Schley [Carya illinoensis (Wagenh.) K. Koch] grown in Australia. J. Agric Food Chem, 49(3), 1277-81.

[1]This Chapter contains much that was originally published in the Proceedings of the Northern Nut Growers Association AGM, 2004,under the heading *Functional Foods – A Comparison of Black Walnut with its Better-Known Cousins*. Nutrient data has been verified against more recent analyses. Some of the original analyses in the NNGA paper may not have been correct, especially in relation to chlosterol-lowering effects of nut consumption. They have been recalculated and expressed differently here.

[2]http://www.fda.gov/Food/GuidanceComplianceRegulatoryInformation/GuidanceDocuments/FoodLabelingNutrition/FoodLabelingGuide/ucm064923.htm

[3] http://www.hc-sc.gc.ca/fn-an/consultation/init/man-gest_health_claims-allegations_sante-eng.php

[4] E.g. http://www.encyclopediaofarkansas.net/encyclopedia/entry-detail.aspx?entryID=542

[5]
http://www.lieffcabraser.com/consumer/walnuts.php?gclid=CKSywaf6xKACFQsNDQodQ32naQ

12. Line Index

And now we come to the heart of what much of this book is about, how within the large population of trees of BNP stands, to identify those individuals that contribute most to the bottom line and why they do it. Please note that I am not advocating a shift away from BNP, but to an understanding of how to use it. This builds on the identification of the heaviest bearers through nut range scoring (NRS) and the work we do to calculate NY (ch 5).

The values in Table 12-1 have been calculated for five '3' lines (trees) as an example. Please note that data collection on this portion of the farm only began in 2006, so not all variables were quantified every year. The identifiers show that all these trees were in Field 1 (followed by Row and Position No). Fewer trees from Field 3 made it into the '3' category

On the far right-hand side are the calculated (multi-year) Alternate Bearing Indices, where it can be seen that 1-4-27 and 1-1-17 are the two extremes, the latter so far expressing high alternate bearing (closest to 1). If you find this hard to understand, or want to know how to calculate it from the above data, see Chapter 5. High ABI or ABNY is not desirable, unless the biennial compensation in NY makes up for non-production in the off years. I am not ready to thin out 1-1-17 just yet.

Line (Tree)			2006			2007				2008				
			A S	NY	DBH	NY	SI	DBH	K%	A S	NY	SI	DBH	ABNY
1	1	14	1	0.05	11.5	3.51	0.69	13.5	21.9	3	3.27	0.70	15.4	0.50
1	2	12	1	3.02	15.0	0.65	0.61	16.7	25.7	1	7.88	0.51	18.5	0.75
1	1	17	1	0.21	10.9	1.81	0.69	12.5	19.5	3	0.00	0.70	14.7	0.90
1	7	12	2	0.22	8.0	5.49	0.67	8.6	21.8	4	2.93	0.70	10.1	0.61
1	4	27	3	1.75	8.0	1.49	0.57	10.0	24.2	4	3.44	0.55	11.3	0.24

Table 12-1. Example of five Line Index entries.

What is the Line Index? It is my quantitative description of the best nut producers using the variables I currently think the most important:

- Anthracnose Score (AS)

- Nut Yield (NY)
- Shell Index (SI)
- Kernel Percentage (K%)

- And, derived from NY, the Alternate Bearing Index (ABI or ABNY)

Below is an explanation of each variable, but not the reason why I consider them the most important. Just be content for the moment that a range of quantitative descriptors has been attached to each line (line = a single tree, which may or may not have an identifiable maternal parent on or off-farm). DBH (Diameter at Breast Height) is included for interpretation.

- Anthracnose Score (1-5). The expressed incidence of leaf-spotting disease (which may have non-anthracnose components), which is commonly considered to depress a tree's productive ability. 1=low, 5=high, thus a 1 indicates nill expression, which can be considered a strong tolerance (if not outright resistance) to leaf-spotting disease.

- Nut Yield (no.nuts per cm² of trunk diameter at DBH). A measure of productivity based on nut count per tree, adjusted for the tree's size. All trees in a plantation will vary in growth rate.

- Shell Index (shell wt proportional to nut volume: g per cm³). A measure of the energy partition occurring at the individual nut level.

- Kernel Percentage (kernel weight as a proportion of dry nut weight). A measure of the edible yield of the nut. Can be used in combination with NY to determine Kernel Yield (KY).

- Alternate Bearing Index (0-1). A measure of the consistency in nut production. Black walnut shows variable biennial productivity. Here adjusted for DBH, thus accounting for tree size, and calculated from NY, thus more truly ABNY, not ABI which, used by others, does not account for change in potential productivity across time based on tree growth.

- Diameter at Breast Height (cm). A standard, convenient measure used to determine annual change in tree size at 135cm from the ground. DBH, for all its convenience, is a moving target. While it remains constant in relation to the ground, it does not remain constant in relation to the top of the tree. Thus, successive annual measures of DBH are taken at different points within a tree's geometry.

Once these variables have been identified ('chosen') we then have to decide how they will be combined for ranking purposes in the Index. The answer to this is that at the moment, we can't. There is not enough information across all variables for a sufficient number of years. But we can begin by working our way towards a solution, and building in improvements as we go.

It is important to remember that what we are trying to do is identify superior trees. The key question is 'Which tree is best?" We assume here that we are referring to nutting ability. But all trees will tend to produce more nuts as they grow larger, so we must adjust our key values for tree size (above all). Hence, the critical importance of NY. Ideally this should be calculated from nut weight produced by a tree, but nut counts are far easier and while they don't account for variability in nut size between trees, they eliminate many other factors (moisture content, etc) which could influence our calculations.

Our purpose in identifying superior trees is to decide what to do with each one (Keep it? Thin it out? See the earlier discussion of thinning strategy) as well as offering an objective means of deciding which nuts to collect and offer for sale (BNP kits?). Too many black walnut 'varieties' are offered for sale in horticultural nurseries without any guarantees even of trueness to type (name) let alone with a description of what the appended name means. While we cannot guarantee that our nuts will replicate the mother tree (the 50% paternal genetics is completely unknown), we will eventually know the partial heritability of the variables we have measured (the degree to which they are consistent from mother to daughter) - we have the second generation of many of these index lines growing on the farm.

It sounds as though I am suggesting a move away from BNP towards an orchard model. I would only say that progress in BNP requires identifying superior trees, and that there is no reason that new, 'improved' BNP stands should not be established from the superior trees of older ones.

As NY appears in our list of variables, we do not want to repeat it on the left-hand side of any equation we develop. In the end I believe that on the left-hand side we put the area based variable relating to kernel yield (kernel yield per unit area; the final economic evaluator of biological issues) and on the right hand side we put all of the above variables plus one (for planting density; were all our trees at 6 x 6 m we could ignore this). We would then test which single variable or combination of variables best predicts kernel yield per unit area across the superior trees (we have already decided on partial superiority by assigning them an NRS of '3'). The way to do this is by something called stepwise regression analysis, which, at the moment, does not require explanation.

Of course, what we are trying to do is predict how trees will perform in the future based on what they have done up till now. We have a certain level of expectation that past performance can be codified, through our variables, in a way that gives us enough confidence that we can do so. Then can we do any ranking at present? And should we? Well, yes, we always want to be testing our methods. Let's say that we want to generate hypotheses that our regression analysis will eventually prove or disprove. Can we, based on the little data we have, make some tentative empirical observations? Not really, as I noted above. Remember, we are dealing with a long-lived perennial species, and should probably have a minimum of five years complete data before we start to rank individuals. I continue to put my faith in NY as the key measure of individual contribution to the bottom line.

The full Line Index as of 2008 can be found in Appendix A.

13. Net Income from nuts

A function of all costs and any other steps not considered

Finally, I have developed a basic spreadsheet which assembles all the key strategic and financial data in a single model that allows us to change some key values and look at their effect on a long-term financial indicator, the internal rate of return (IRR), which is really how we should look at net income, i.e. a stream of annual values which tell us whether we can make a profit from nuts. By definition the IRR is a measure of the profitability of an investment, and is often used in comparative analyses. My spreadsheet can be downloaded free[1], which gets around the enormous difficulty of printing spreadsheet output. You can also see how calculations are made. It is comprised of five components, some of which correspond to certain chapters in this book. After the initial explanation I run some sensitivity analyses for you to see the relative change in IRR according to the changes made. It is this relative change which is important, not the absolute change, as relative change will be less sensitive to the assumptions made at the beginning[2]. With the current/base values, IRR is estimated over 10 years at 9%.

Component 1 – Fruit composition

This component allows us to calculate a corresponding kernel value if we were to buy in fruit at a given cost. This is standard business practice in certain agricultural industries where it is wanted to know how much each stage is contributing to the bottom line. Thus, a vineyard has to value the grapes it sends to its winery, and so on. We need to know the cost of processing so that we know what value addition occurs at each step in the chain.

I have introduced factors for fruit composition which are intended to highlight the relatively low contribution of the kernel to overall fruit weight, and thus its high intrinsic value at any fruit price. By varying the component percentages and the buy-in price you can also see that a buyer would be restricted to ¢/kg if the kernel value were not to sky-rocket beyond a point where later inefficiencies became problematic.

Component 2 – Processing rates

This takes each step of our processing system and applies flow rates so that we can see where the bottlenecks occur. This is important to later calculations.

Component 3 - Market values

This is where we insert what we believe we can charge the wholesale and retail markets for our products. Only three products are currently mentioned and one of them (hull) currently has no value assigned.

Component 4 – Source

My own experience has shown me that I need to balance on-farm production with off-farm sources from a different ecological zone. I have labeled the farm *zone 1* and have included two sources from other zones. The *zone 3* source works on the Hammons model, where I assume I have an almost zero-cost labour source who will deliver nuts from the surrounding landscape and buy the resulting kernel from me at a wholesale price for their own fund-raising purposes. Here I acknowledge that the trees in all zones are older than mine and likely to be more productive.

Component 5 – Business summary

This component takes values from the preceding components and assembles them in a way that allows us to summarize costs and returns. The returns-to-investment section takes a growth factor for yield per tree and applies it to the yield data at the chosen growth rate for calculating returns. The cost assumptions are to be found at the foot of the table. Machinery investment is entered as a fixed cost and this line is where adjustments can be made to reflect real cost and rate of investment.

Sensitivity analyses

Several sensitivity analyses show that the bottom line can be significantly affected by changes in a variety of factors. These boxes are both at the right of the spreadsheet and reproduced here. Increasing kernel percentage by 5% doubles our IRR, suggesting that if we'd started with better selections we'd be further along, but at this

Sensitivity analysis 1	
Changing kernel %	
New factor	25%
IRR	20%

110

point we still don't know what changes in other factors will do, so let's not be hasty. Throughout these analyses we assume complete efficiency in separation, though this is questionable. What is probably true is that this efficiency should not change with change in other factors.

Sensitivity analysis 2	
Changing separation rate	
kg/day	
New factor	12
IRR	45%

Changing yield per tree across the whole spectrum of zones brings the largest change, though elevates our IRR into stratospheric relative values which probably have only some basis in reality. We can capture some of this increase by being successful in doubling on-farm yield/tree. A change in our kernel separation rate by a factor of 2 brings another large relative change in IRR. Kernel separation rate needs to influence annual kernel sales for it to have an effect.

Sensitivity analysis 3	
Changing yield/tree	
New average 5kg	
New factor	5
IRR	50%

These are not analyses upon which necessarily to base management decisions, but rather cases which show that our enterprise is subject to a wide variety of influences, of which yield per tree is an obvious one. A doubling of number of trees on-farm is as effective as doubling yield/tree, adding weight to our BNP concept.

Sensitivity analysis 4	
Changing yield/tree on farm	
New average 2kg	
New factor	2
IRR	20%

Experience will allow us to enter realistic value for each of the factors but still allow us to test the effect of other efficiencies and so suggesting where we need to concentrate. Obviously we are not going to change our total number of trees after 20 years if we started off with the number we thought we could manage, though in our case we have a graduated planting pattern that will bring other trees 'on-line' as time goes by. Some of these are selections from within our own stands.

Sensitivity analysis 5	
Changing no of trees on farm	
New factor	3000
IRR	20%

Increasing the rate of investment in machinery, either because of real cost or what we need our processing line to look like in the early years, reduces IRR, but only by 3% if investment costs were 50% higher in the first five years.

Sensitivity analysis 6	
Machine investment	
New factor	50% increase
IRR	6%

Summary

This model suggests that at the deliberately conservative values for factors I have chosen, there is money to be made in BNP. Every improvement in factor values brings an increase in relative IRR, though it is perhaps the early income years which most affect our outcome and where we would want to concentrate our improvements. Higher separation rates coupled with increased yields per tree (or numbers of trees) would be very effective in improving the bottom line. Investment costs appear less important than stand-management efficiencies in their effect on IRR.

---

[1] From http://blackwalnutsdotca.wikispaces.com/

[2] As highlighted earlier in Ch 2, alternative models exist, e.g. http://www.centerforagroforestry.org/profit/walnutfinancialmodel.asp though the range and nature of assumptions in this alternative case are quite different from those in the spreadsheet discussed in the present chapter.

14. Afterword

In the first chapter I noted three (of possibly many) questions that could be asked by a fence-leaner:

- My purpose is making money. Can I make more money from that than from what I'm currently doing?
- I have this land that I'm not using. Will black walnut for nuts bring an appreciation in value?
- I want to grow nut trees. What should I plant and how should I do it?

I think these questions have been answered, respectively, to varying degrees:

- I am fairly certain you can make money, though how much you can make will depend on the originality and efficiency of your business model;
- If other people value a nutstand as you do. Which depends on being successful in your original business premise, and then communicating that fact;
- Ask someone else, but if BNP appeals to you, then follow the black walnut path as outlined in this book.

I have endeavoured to communicate what we currently know, and make the scientific basis for this knowledge as interesting as possible. Knowledge of this nature cannot be generated without science. If I have missed things, or made mistaken conclusions in some places, time should eventually correct these. But I had to start somewhere.

I haven't answered the timber question, primarily because I have no evidence, and I am trying to concentrate on doing one thing well. Everything so far says forget about long-term timber goals and focus on the immediate revenue streams. The two don't mix. Look on the Internet for a picture of a California walnut grove if you think otherwise.

We are on the edge of a revolution. Of agriculture (from a natural-resource management perspective); of nutrition (healthy and health-giving diets); of rural-economies (how to continue to generate a partial livelihood in the most sustainable way possible); of landscapes (rural ones principally, as a result of all these aspects). Leading a revolution is the only way to feel in control of it. This depends on a confident vision of what is possible.

Finally let me say that I enjoy living surrounded by trees, and that if you think I'm a bit eccentric in my views, so be it. Life is made interesting by the tangents that arise. We can get back to a multiple-income stream inter-generational model that can provide families with the means to stay on the land, and for future generations to add to leaving it in a better condition than their forebears did.

I hope the journey has been interesting.

Appendix:A

Full Line Index
of Superior Trees
2006-08

Line (Tree)			2006			2007			
			DRS*	NY	DBH	NY	SI	DBH	K%
1	1	5	3	2.29	8.20	1.82		9.20	
1	1	10	3	5.34	8.20	4.24		9.20	
1	1	11	4	0.42	10.10	0.32		11.70	
1	1	14	1	0.05	11.50	0.03	0.69	13.50	21.85
1	1	15	2	0.56	10.80	0.46	0.67	11.90	19.29
1	1	17	1	0.21	10.90	0.16	0.69	12.50	19.47
1	1	21	3	0.16	7.40	0.11	0.94	9.10	22.11
1	1	22	3	1.27	8.20	0.05	0.65	9.00	22.76
1	1	23	2	1.35	10.70	0.97		12.60	
1	1	28	3	0.12	10.70	0.10	0.66	8.10	21.40
1	2	4	4	7.72	9.80	6.35		10.80	
1	2	5	3	1.36	10.20	1.21	0.69	10.80	21.60
1	2	6	4	3.40	10.00	3.03		10.60	
1	2	8	3	1.50	8.30	1.07		9.80	
1	2	9	2	2.75	15.00	3.07	0.70	14.20	21.27
1	2	10	3	1.82	13.60	1.40		15.50	
1	2	11	4	4.42	8.80	3.43		10.00	
1	2	12	1	3.12	15.00	2.44	0.61	16.70	25.67
1	2	13	3	0.10	7.20	0.07		8.50	
1	2	15	4	0.72	8.70	0.65		9.20	
1	2	17	3	0.00	6.10	0.00		7.70	
1	2	18	4	0.60	7.00	0.51		7.60	
1	3	3	2	1.21	7.60	0.95	0.61	8.60	22.87
1	3	12	3	5.80	9.10	4.23	0.62	10.70	26.79
1	4	8	3	3.41	7.70	2.56	0.64	8.90	24.95
1	4	17	4	0.00	7.60	0.00		8.30	
1	4	23	3	1.75	8.00	1.12	0.57	10.00	24.20
1	4	27	3	0.81	7.20	0.53		8.90	
1	4	30	3	0.00	9.40	0.00	0.63	10.50	24.04
1	5	5	3	0.68	8.20	0.47	0.73	9.90	17.39
1	6	29	3	1.46	6.40	1.06		7.50	
1	7	11	4	1.37	8.40	1.20	0.71	9.00	22.63
1	7	12	2	0.22	8.00	0.19	0.67	8.60	21.80
1	7	29	4	0.22	6.30	0.00		7.20	
1	9	22	3	0.00	8.40	0.00		10.30	
1	9	23	3	2.31	7.00	1.69	0.62	8.20	28.75
1	10	21	3	1.02	8.50	0.74		10.00	

Line (Tree)			2008				
			DRS*	NY	SI	DBH	ABNY
1	1	5	1	0.68	0.68	10.70	0.29
1	1	10	2	3.20	0.65	10.40	0.13
1	1	11	4	2.55	0.7	13.00	0.46
1	1	14	3	2.52	0.68	15.40	0.61
1	1	15	3	4.23	0.62	13.70	0.45
1	1	17	3	1.94	0.7	14.70	0.49
1	1	21	2	0.23		11.20	0.27
1	1	22	2	2.90		10.60	0.95
1	1	23	1	3.73	0.73	14.00	0.38
1	1	28	5	1.48		9.00	0.48
1	2	4	4	5.94	0.65	12.40	0.07
1	2	5	4	2.43	0.67	12.40	0.20
1	2	6	4	3.97	0.59	13.10	0.10
1	2	8	3	3.25	0.62	11.70	0.34
1	2	9	2	3.42	0.73	16.90	0.05
1	2	10	4	1.94	0.68	17.50	0.15
1	2	11	2	3.96	0.74	11.30	0.10
1	2	12	1	6.42	0.51	18.50	0.29
1	2	13	3	3.32	0.6	10.90	0.57
1	2	15	4	4.77	0.71	10.40	0.41
1	2	17	3	4.99	0.44	8.10	1.00
1	2	18	4	5.27	0.69	9.10	0.45
1	3	3	2	3.20		10.50	0.33
1	3	12	3	6.37	0.59	12.50	0.18
1	4	8	4	9.79	0.55	10.00	0.36
1	4	17	3	4.03	0.65	9.50	1.00
1	4	23	3	2.69	0.55	11.30	0.32
1	4	27	4	4.24	0.61	11.00	0.49
1	4	30	3	2.64		11.10	1.00
1	5	5	4	0.53		10.90	0.12
1	6	29	5	4.07	0.63	8.20	0.37
1	7	11	2	0.79		11.20	0.14
1	7	12	4	2.12	0.7	10.10	0.45
1	7	29	4	1.73	0.74	8.70	1.00
1	9	22	2	4.26	0.62	11.50	1.00
1	9	23	4	0.44		9.60	0.37
1	10	21	3	1.94	0.51	11.10	0.30

Line (Tree)			2006			2007			
			DRS*	NY	DBH	NY	SI	DBH	K%
3	5	30	3	0.70	6.60	0.48	0.62	8.00	25.07
3	6	12	2	3.20	8.70	3.05		8.90	
3	7	14	1	0.67	8.50	0.74	0.63	9.20	22.77
3	7	26	4	3.09	8.10	0.00		9.50	
3	8	15	3	0.01	7.70	0.40		9.90	
3	8	16	2	0.25	7.20	0.04	0.63	9.30	19.53
3	9	18	4	0.01	5.00	0.08		5.50	
3	8	38	3	0.00	5.00	0.00		6.00	
3	10	13	3	4.65	5.90	0.00		6.60	
3	10	28	4	0.00	3.20	0.00		4.20	
3	10	34	3	0.00	6.10	0.12		7.70	
3	10	37	4	0.03	4.20	0.28		5.20	
3	12	27	2	0.25	3.20	0.19	0.66	3.70	25.65
3	13	33	3	0.25	6.80	0.17	0.69	7.20	20.60
Averages			42.27	1.42	8.24	40.37	0.67	9.33	22.72

Averages calculated from no.of observations for both fields for any year's parameter

*DRS values from 2005 and 2007 for Field 3 (3-- Lines) only

Line (Tree)			2008				
			DRS*	NY	SI	DBH	ABNY
3	5	30	3	0.95		9.20	0.26
3	6	12	2	1.87		11.00	0.13
3	7	14	1	0.49		11.70	0.13
3	7	26	4	2.25		11.00	1.00
3	8	15	3	2.23	0.66	11.40	0.82
3	8	16	2	0.11		11.50	0.60
3	9	18	4	4.06		6.60	0.87
3	8	38	3	2.56		8.50	1.00
3	10	13	3	2.92		8.00	1.00
3	10	28	4	4.98		6.00	1.00
3	10	34	3	4.53	0.68	8.40	0.97
3	10	37	4	11.74		6.10	0.88
3	12	27	2	4.06		3.80	0.52
3	13	33	3	0.88		9.60	0.43
Averages			42.41	3.25	0.64	10.88	0.50

Appendix B. As I walk the walnuts, from the Biomastery blog

As I walk the walnuts - 1

As I walk the walnuts, which I do twice a day when I'm on the farm, Kahlua with her nose in the grass somewhere, I remember comments from some visitors who seem offended by my approach to tree management: *you really should get rid of those lower branches*. Not everyone tells me why – some assume that I'll get the message that they know better than I do. Others clearly come from a forestry background, where any branch on the first 30' of bole is not only an eyesore but probably also an offence punishable by excommunication from the College of Foresters. None actually asks me why I've left so many branches.

My purpose is actually to let the trees fill their space. Removal of a branch cannot be undone. Branches are a tree's contact with its environment. Remove a branch and you have reduced that contact. So what?

The conical growth rule discussed earlier 'builds' [see a later chapter] on one simple principal, relative growth, i.e. that growth is a consequence of growth accumulated before. Remove some of that growth and you immediately reduce the tree's future potential growth. It affects the leaf area the tree can subtend, and the surface area upon which it can lay down the present year's captured carbon. An open- grown tree explores every opportunity to push its tendrils out into unoccupied territory, and if I leave those lower branches where they are that territory is at my height, where I can see and feel the tree's features, its health, and thus , in purely practical terms, its productivity.

Biomastery blog post, Nov 21, 2007

As I walk the walnuts, and as there are about 2,500 of them it can take quite a while, a discussion ensues: morning wally, g'day wally, looking good wally, etc., etc. My wife would be happier were she to know that no words are actually spoken, but the occasional murmur after visual inspection is unavoidable. Kahlua occasionally intrudes, with notice that she has uncovered something edible, but generally it is a peaceful intrusion, a thrust of wet nose into the hand. She knows my path, and will meet me at the end of it.

So, to answer your unspoken question, do I prune the trees at all? Well, yes, but probably not in the way you think. Of the 2,500, there are about 1,800 that receive an inspection, generally in March, with a rapid appraisal of growth and form, and perhaps a cut here and a cut there. All inspection is from the top down, with most attention paid to serious clefts which I believe could prejudice future strength. I have learned not to worry too much about lack of obvious leaders. The tree has its own means of assigning leadership among its myriad of branches, and it is rare to find instances of consistent competing dominance which could seriously deform a tree. If there is a cleft I don't like, it is a two-year job to remove it, to avoid a major wound before carbohydrate resources have been assigned within the tree and increased relative deposition rate on the full branch remaining. Regular cuts are made above the collar, and the cut surface trimmed so that as the new cambial layer slowly engulfs the wound it does not have to negotiate right-angled turns. A trimmed cut will be covered within half the time, or even less (thank you, Malcolm Olsen).

The remaining 700 are still small enough that pruning is a process of early formation, to avoid problems that would make correction more difficult later. But even then a tree is best left alone, with only minor corrections. Splinting or taping to correct obtuse forks is the only exception.

Stay away from bottom-up pruning. It would satisfy the College of Foresters, but it will leave you with fewer future options, reduced growth, and a tendency not to address higher canopy concerns. Let the tree fill all the space it can.

Biomastery blog post, Dec 4, 2007

As I walk the walnuts, or more accurately, pass through on snowshoes, as we had about 25cm of the white stuff on Friday, I look ahead to the coming season and wonder what's in store for us. Maple syrup production has turned into a big gamble, because of early sap flow when the temperature spikes, and a friend of mine wonders whether it will ever be worth tapping the trees again. He's slowly building a new sugarhouse, in case he becomes so inclined, but it doesn't look as though it will be worth his time hurrying this year. The corollary for a nut producer is those aggravating late frosts in May, which can sweep the trees clean of the emerging flowers. So far, the walnuts will take 2°C up until the end of the first week of May, after which it is a complete gamble. Last year, a frost on May 19th did widespread damage, though it didn't come close to the two consecutive nights of -5 °C we recorded about the 8th May in 2005.

Common sense says we need to track the phenology of flowering, because later emerging flowers stand a better chance of missing the frost. However, this is a laborious task, which would require scoring all our trees. Well, you say, you already score for disease expression and nut production, why not phenology? Well, I say, if a tree bears nuts after a frost in the critical period I mentioned, there's a good chance it did so because it flowers late, i.e. I'd rather score the result than the cause. However, now that I have this indicator, it would be a good idea to track the phenology of flowering on the 25 trees from which I on-planted last fall. My hypothesis is that all valuable traits have at least some measurable heritability from the maternal line, and that I need to determine whether the average expression of each trait in the population of offspring is at least marginally better than that same trait expressed in the maternal tree, i.e. that I can make genetic gain by selecting under our conditions for nut yield, kernel quality or whatever other characteristic I believe valuable. However, it takes that first effort at on-farm selection to know whether it is worthwhile (equivalent to concentrating beneficial characteristics in a smaller population than the one from which the selections originated).

Biomastery blog post, Feb 3, 2008

As I Walk the Walnuts – 4

As I walk the walnuts, I reflect on the changes in bird life on the farm. Back in the early days, before the trees it was pretty sparse, though there were the annual visits from bobolinks, snow buntings, and red-winged blackbirds. Even the occasional kestrel, though this was a function of the hydro cable crossing the farm, and thus the prior existence of a superb scouting perch.

Those all still come, though in ways they didn't before. The open areas of standing grass between the trees continue to attract the ground-nesters, and the trees themselves offer nesting sites for birds that wouldn't have nested before. The red-winged blackbirds like to conduct their mating rituals on the aerial infrastructure the trees provide, making a good racket in the process. The kestrels are as likely to perch on the trees now, and it is wonderful to watch their young in their cart-wheeling antics up and down the rows in fall. This year, the snow-buntings may have had a harder time finding food in the open fields because a flock of a hundred or so took up semi-permanent residence in the few trees just outside the kitchen window, thus close to the bird feeder, and would come swooping in every half-hour or so, like a school of tropical fish flashing over a reef, all taking some sort of signal from one, feeding as a flock and not as individuals, taking flight together when someone called the time out. On those cold winter days these were my family, and I took great care to look out every hour or so, to make sure that there was sufficient seed for them, on the ground as much as on the table, as most would feed from that swept off the table either by their mates or by the blue jays that come in and shovel around, looking for the elusive peanut amongst all that other stuff.

Biomastery blog post, Mar 23, 2008

As I walk the walnuts – 5

As I walk the walnuts, either measuring pole or diameter calipers in hand, as it is measuring season, there are things that I notice. It is spring here, though the trees are one of the last to emerge from their winter solitude. When I am measuring height, I often look beyond the tip of the tree to a solitary bird well up in the sky making use of the currents that there must be at that altitude. Not geese, for they announce their passing well in advance, and clear the sky by force. No, these are occasional hawks or gulls that catch my eye and make me pause and wish that I had their talents for silent passage. Where are they going, and who awaits them?

When I am measuring diameter (I measure height and diameter separately, because I struggle to carry all those bits and pieces at one time) my gaze is closer to the ground. Why, I wonder, is there an apparent association between ants and black walnut trees? In my field where I collect most of this data, there is commonly one ant hill between 50cm to 1m from the base of the tree, in the grass-free band along the rows of trees. I frequently find the ants cruising the tree, though sometimes stationary on the terminal buds, and I can only surmise that they are feeding on the sap I sometimes see escaping from larger wounds. But I am amazed that the rule is one hill per tree, where I find them (not all trees have anthills). If this were not a deliberate association, I would expect to find anthills at random distances between pairs of trees, and perhaps in multiples, but this is not the case. Who passes the message on? This is my tree, mate, go and find your own (actually 'ours' given that ants are social insects, but 'ours' doesn't mean 'yours too'). Perhaps there is a scent trail at the base of the tree which says just that: Trespass at your peril!

Biomastery blog post, Apr 27, 2008

My tree population has become a world unto itself, massive growth this year, uplifting (in all senses) avian biodiversity from the relative paucity of two-dimensional flatland of the original hayfields into the third dimension of biomass infrastructure – that fractal takeover of the near-sky by carbonic tentacles that are uniquely black walnut.

It is when I stand in these 6m high avenues that I am aware of something fundamentally clear - that the immediate biosphere is recovering after some 150 years of agricultural exploitation and that the animate occupants of this space are the more joyful for it. Of course, they are instinctively animated, so the joy is nothing more than behavioural change or enhanced presence because of changing predator-prey dynamics, but I prefer to put an anthropomorphic slant on things. After all, I am more joyful, so why shouldn't they be?

Elsewhere in this blog I have explained what I mean about biomass walnut production, and my 'non-cultivar' approach. As the years pass I have become more aware of what this broad diversity means to the farm landscape, and how I believe it adds far more value than would planting reduced numbers of named selections (the dollar-for dollar-implication). In scientific terms, I have gone back to a genetic baseline, providing the means to identify a benchmark against which to detect change; in spiritual terms, I have given worth to individuals in my populations on terms quite distinct from the do-or-die dogma of modern agriculture. At the risk of being thought to have gone over to a lunatic fringe, it important that I expand on this. To do this, I'll bring Thomas Merton into the discussion.

Thomas Merton was a Cistercian monk who deliberated long and hard on Nature. Because he came to it as a God-fearing person, he spent much of his time contemplating its sacredness. By the definition of 'sacred', Nature was made holy by religious association. As a holy person (said by others), he was therefore unlikely to consider Nature was non-holy. Merton died in Bangkok in 1968, a few short years before I got there, but I have only really discovered his writings recently (unlike John Stewart Collis, subject of an earlier post). Merton was also a bit of a blogger, and I have quite enjoyed his When the Trees Say Nothing (Sorin Books), but if I have an argument with him, and other writers, e.g. Ursula Goodenough (The Sacred Depths of Nature, Oxford University Press, 1998) who depart from the station by the same platform, it is that they are almost certain to confuse holy (to be revered) with holy (belonging to or empowered by God). I believe that holiness can exist without referencing a deity, and that the reverence I feel when I stand in an avenue of black walnuts derives from their capacity to fill my vision in a particular way, both when still and when moving, and that they now dwarf me, returning me to the natural condition when humans evolved amongst the trees. Perhaps that reverence has a hint from genetic memory of the dangers associated with tall trees. Otherwise, it may rest in the pleasant reflection that trees now exist where they haven't for 150 years.

But, you see, Merton is right about those infinite possibilities, though they are as much to do with not knowing much about the future, and enjoying ourselves responsibly on the road Under Nature as much as possible. In a climate undergoing rapid change, I would rather be caring for thousands of not very expensive trees, than losing sleep at night because of what I have invested in a few. Many of my trees may not express worth in the way it is commonly measured today, but until I come to measure it I'll enjoy having them around.

Part of Biomastery blog post, July 24, 2008.

As I walk the walnuts – 7

As I walk the walnuts, !st January 2009, -18°C, I think about the year to come and how much farther we will get. First, though, Happy New Year to all my collaborators, from the tree end to the machinery end. None of this would work without you. To warm you up on a cold winter's day, I have chosen a summer photo [on the blog only].

Then, a swing past the barn, where, a short six weeks ago, I was finishing up the wet processing of the 2008 harvest. My rapid departure to warmer climes for a couple of work weeks is evident: the final pile of hulls is still there, along with the mess made by three squirrels I have recently managed to move out. The final nuts are still on the dryer, though the remainder are bagged up and labeled ready for weighing.

In the fields, the range scoring tape is evident, splashes of colour on the naked trees. How much that system has helped me to understand what is going on at the tree end. It will come off at pruning time to avoid confusing next summer's work, when the trees will be re-scored.

Still to do is the nut report, which entails cracking and measuring the kernel and shell fractions of the '3' trees. That will happen sometime in February. Observation of the samples suggested some surprising things to come, but I will wait for the hard data before saying anything more.

So, 2009 will be the year of 'nutmastery' and a separate blog to deal solely with that end of things – machinery, markets, etc., while this one may slowly wind down. We'll see.

Biomastery blog post, Jan 1, 2009

As I walk the walnuts – 8

As I walk the walnuts and look at each tree, it is like a snapshot. Snapshots, by definition, are images captured at an instant in time, and tend to reflect something that we were doing at that given moment. Yet an image of a tree is misleading. It is certainly a snapshot of the tree at that given moment, but the tree is the ultimate interpolator, the summer-up of expression over its lifetime, the presenter of the all-gone-before. A tree, then, while in the present, is all of the past, as interpreted by its individual genome in that unique environment.

I say this because a tree is not just the prisoner of its ability to fix and deposit carbon according to a given pattern, but because, once started, it cannot deviate from this roadmap. Oh, the roadmap may be altered for it – spring frost effects on bud growth, branch removal by pruning - but a black walnut always looks like a black walnut, no matter its shape, and will always yield nuts, not acorns.

With time I am less involved with the form my trees take. I matched their early growth with my own energy, interfering according to my criteria at the time, but now am content to let the trees fill their space however they will, their response to that interference an acknowledgement of my then-presence on the landscape, an element of biodiversity as ephemeral perhaps as the guarantee that the snow buntings will return every year. Nothing is guaranteed, not even that the energy to interfere goes undiminished, but I prefer to think that I just understand myself better now, and that however the trees fill that space is the way it should be.

Biomastery blog post, Mar 1, 2010

As I walk the walnuts – 9

As I walk the walnuts I think about the graying of the population and the fact that the mean age of the attendees at past nut society meetings has been greater than my own, and I am on the cusp of retiring! The fundamental challenge is how to interest young people. The answer, I believe, is only through income potential. Remember that question? Can I make more money from that than from what I'm currently doing?

But perhaps there is another way. And that is to build associations in the mind between nuts and play. To that end, behold the Walnut Express, a model railroad layout that snakes around a peripheral shelf in the cracking shed, ferrying nuts from the Walnut Mountain (or Forest!) to the cracking machine. It's still under construction so I am not sure whether it's a Mountain or a Forest, and only time will tell. But one thing of which I am certain – the cracking shed will be a boring place for young people without it. Just ask my grandchildren.

Biomastery blog post, Mar 23, 2010

As I walk the walnuts – 10

As I walk the walnuts I reflect on the progress we have made in the last five years and my hopes for this one. All of the technological components are beginning to come together, albeit at different rates, and in spite of glimpses of future bottlenecks that seem to parallel what our climate throws at us, adding to the challenges of nut production so far north, we have the market waiting for what we can supply, so no fears there.

Useful US studies show that we import about $110 m annually of tree nut products from the US alone (the study was commissioned to find this out). I raise gales of laughter when I tell my collaborators that I'd be happy with 1% of that market, but the truth is that if we could achieve a gross return of $25,000 annually to 40 landowners, we'd be well on the road to a sustainable partial livelihood across the region, which is far preferable to one or two producers making greater gains.

But this still requires effort, and perhaps more than we have brought to the task so far. I shall be egging my collaborators on even more, trying to increase the range of skills we bring to the questions that remain, showing why the biological issues of productivity parallel the technological challenges of bringing a product to market. I am more convinced than ever that BNP was, happily, a viable strategy, and that it is easily replicable on a wider scale. But there is a lot of black walnut currently out there on the landscape, and we need to harness this resource, as well, to add to the B of BNP and to our bottom line.

I see as a challenge now, how to add a focus on shell usage as a biofuel. With one of the highest energy densities in natural by-products, and at about 75% of our physical output, shell can be (should be) part of our market strategy, and while I am aware of its usage as an abrasive, amongst other things, in our climate this energy density cries out for a different end-use. We are in the age of renewable energy. Not exploring this would be like throwing the baby out with the bath-water.

Biomastery blog post, Mar 30, 2010

www.ingramcontent.com/pod-product-compliance
Lightning Source LLC
Chambersburg PA
CBHW020208200326
41521CB00005BA/289